PRIMITIVE NERVOUS SYSTEMS

PRIMITIVE
NERVOUS
SYSTEMS

THOMAS L. LENTZ

New Haven and London
Yale University Press
1968

TO J. P. L. AND S. T. L.

PREFACE

The nervous systems of lower invertebrates, although capable of carrying out some complex functions, are generally less specialized than the nervous systems of higher invertebrates and vertebrates. Study of these primitive nervous systems may be more likely to reveal the basic or elemental properties of nervous tissue and the characteristics of the earliest nervous system to appear during evolution. It has become apparent, however, that the older theories on the origin of the nervous system cannot entirely account for the more recently discovered facts of neurophysiology and neuron structure in these lower animals. It is necessary also, in considering the origin of the nervous system, to incorporate the findings of all areas of neural investigation, including biochemistry, morphology, neurophysiology, and behavior. For these reasons the neural characteristics of selected lower invertebrates are surveyed in this book, and some suggestions are made concerning the nature of the primitive nervous system, its origin, and its subsequent evolution. Despite recent advances, many aspects of primitive nervous systems are poorly understood or have not been investigated. As a result, discussions of the origin of the nervous system must be speculative in some respects. It is hoped, however, that the considerations of this topic will at least point out those areas where knowledge is deficient, thereby suggesting additional experimentation and study.

This book reviews the present state of knowledge of the nervous systems of some lower animals. The nature of the simple metazoan in which nerve cells may have first appeared is reconstructed, and some of the earlier theories on the origin of the nervous system are summarized. After this introduction the current experimental data concerning the neural organization of three primitive animals—sponges, hydra, and planaria—each representing different levels of complexity, are presented. The final chapter, on the basis of this information, advances a hypothesis to explain the origin and probable characteristics of the earliest nervous system to appear.

I would like to thank Drs. Russell J. Barrnett, Floyd E. Bloom, Thomas R. Forbes, and William U. Gardner for reading portions of the manuscript and making helpful criticisms and suggestions. The cooperation of the publisher and especially the editorial work of Mrs. Anne Wilde of the Yale University Press is greatly appreciated. I am also indebted to a number of my colleagues for their assistance, advice, and encouragement while my own investigations were in progress. My research was supported by grant TICA 5055 from the National Cancer Institute, National Institutes of Health, United States Public Health Service.

Thomas L. Lentz

New Haven, Connecticut
1967

CONTENTS

Preface   vii

1. Primitive Nervous Systems and Theories of Origin   1

2. The Neural Organization of Sponges   14

3. The Nervous System of Hydra   36

4. The Nervous System of Planaria   69

5. The Evolutionary Origin of the Nervous System   103

References   125

Author Index   139

Subject Index   143

ILLUSTRATIONS

Figure

1.	Histology of *Sycon*	17
2.	Epinephrine and norepinephrine staining in *Sycon*	25
3.	Granule-containing mesenchymal cell in *Sycon*	29
4.	Granule-containing mesenchymal cell in *Sycon*	33
5.	Histology of hydra	39
6.	Methylene blue staining of the nervous system of hydra	47
7.	Ganglion cell of hydra	51
8.	Neurosecretory cell of hydra	55
9.	Sensory cell of hydra	59
10.	Neurosensory cell of hydra	63
11.	Nerve endings in hydra	67
12.	Nervous system of a planarian	73
13.	Peripheral nerve plexuses of a planarian	77
14.	Acetylcholinesterase localization in a planarian	81
15.	Ganglion cell of a planarian	83
16.	Nerve cell in the brain of *Dugesia*	87
17.	Granule-containing nerve cell of a planarian	89
18.	Granule-containing nerve cell of a planarian	93
19.	Neurosecretory cell of a planarian	95
20.	Neuropil of the brain of a planarian	99
21.	Photoreceptor of a planarian	101
22.	The primitive nervous system	105

Table

1.	Neurohormones identified in lower organisms	112–13

1

PRIMITIVE NERVOUS SYSTEMS
AND THEORIES OF ORIGIN

Introduction

The evolutionary origin of the nervous system has been of recurrent interest since the latter portion of the last century, and this interest has been stimulated by a desire to reduce the extreme complexity of the vertebrate nervous system to its basic or elementary properties. By this means it might be possible to uncover the underlying mechanisms—chemical, morphological, and physiological—of nervous function.

Speculation on the origin of the nervous system is often an attempt to reconstruct an elementary nervous system, which can be described as a group of nerve cells with the minimal number of specializations required to perform the basic functions of nervous tissue. It is important, however, not to confuse the conceptual model of an elementary nervous system with the actual characteristics of an early, simple system, since these are not necessarily the same. For this reason, the earliest nervous system to appear is referred to here as the primitive nervous system.

Comparative physiology and anatomy provide most of the evidence for determining the nature of the primitive system and its origin. The nervous systems of lower invertebrates are also primitive in that they do not have all the specializations present in higher invertebrates and vertebrates. The most primitive metazoan groups include the Porifera, Coelenterata, and Platyhelminthes, and these groups, in particular, are most likely to reveal the characteristics we seek. Sponges, hydra, and planaria have been selected from these phyla as representatives of different levels of primitive nervous organization. From these existing simple forms it may be possible to infer the events that led to the formation of the nervous system. Some caution is necessary in this approach, however, because many of the lower forms have achieved highly specialized and successful adaptations to their environment, which require specialized or complex mechanisms. These nervous systems are less specialized only from a relative standpoint. There are also some striking and conspicuous similarities in the structure and function of nerve cells in widely separated groups of animals. Although some of these characteristics probably evolved independently in different groups, their general and widespread occurrence in some cases suggests a common origin and primitive nature.

No attempt will be made here to determine in terms of evolutionary theory the forces or processes that resulted in the changes leading to formation of the nerve cell. It is assumed that evolution of the nerve cell involved the selection of cellular variations that gave the organism some adaptive advantage expressed in improved chances of survival. Thus, evolution of the nerve cell was probably a gradual process of change in form and function progressing to a level of specialization at which a nerve cell could be said to exist.

Origin of the Metazoa

Most considerations of the origin of the nervous system are based on the neural organization of the coelenterates, the lowest group definitely possessing a nervous system. It is usually recognized that this level of organization must have been preceded by a somewhat simpler system, although the animal form in which the most primitive neurons occurred has not been specified. The nervous system, however, cannot be separated from the behavioral capabilities of the animal, which depend partly on the types of effectors present. Furthermore, in discussing the origin of the nerve cell from a prenervous cell type, it is necessary to know what other cell types were present at the time or before the nervous system appeared.

For these reasons, some knowledge of the structure and behavior of the primitive animal is helpful.

Although sponges are simpler than coelenterates and may have a very primitive type of neural organization, they are not in the direct line of evolution to higher metazoans. Whatever neural organization sponges have is probably not the forerunner of the nervous systems of higher Metazoa, even though in certain respects it may resemble the primitive nervous system. The origins of the nervous system, therefore, should be sought in a primitive metazoan ancestral to the Eumetazoa.

There are two widely differing theories on the origin of the Eumetazoa (see review by Hanson, 1958). One proposes that multicellular animals arose from the Protista by the formation of a colony of individual cells; the second holds that multicellularity developed by cellularization or appearance of cell boundaries in a multinucleate protistan. The first suggestion is the more widely accepted and is based on Haeckel's (1874) gastraea theory which has been modified by Hyman (1940, 1942). According to this view, the Metazoa arose from a hollow spherical flagellated colony resembling *Volvox*. The common ancestor of the coelenterates and flatworms is a planuloid form which arose from the hollow cavity by differentiation of somatic and reproductive cells followed by the appearance of locomotor-perceptive cells and nutritive cells which wandered into the interior. The planuloid ancestor gave rise to the coelenterates and, by way of an acoeloid form, the rest of the Eumetazoa. The Porifera are included in a separate branch of the animal kingdom, the Parazoa, thought to have diverged early from the metazoan main stem.

The planuloid ancestor, resembling the planula larva of coelenterates, is considered to have been elongated, radially symmetrical, and without a mouth (Hyman, 1951). It consisted of an outer ciliated or flagellated epidermis possibly composed of epitheliomuscular cells and a solid interior mass of digestive cells. The outer layer of cells presumably caught food in a protozoan manner and transferred this material to the interior cells which

3

digested it intracellularly. Undifferentiated cells capable of differentiating into gametes and the other cell types also occupied the interior mass. A nerve net was present beneath the epidermis, and sensory cells were concentrated anteriorly.

According to the second theory, the Flagellata gave rise to the Ciliata which are the ancestors of the Eumetazoa (Hadzi, 1953, 1963; deBeer, 1963). The ancestor of the Eumetazoa was a multinucleate ciliated protozoan that evolved into an acoel flatworm, the most primitive eumetazoan. The Acoela gave rise to the Eumetazoa, including the coelenterates. The Parazoa are considered to have evolved from colonies of choanoflagellates.

The Acoela are bilaterally symmetrical, flattened worms. An outer ciliated epidermis* surrounds a solid mesenchyme. A subepidermal and parenchymal musculature is present, and some acoels may possess epitheliomuscular cells. The mesenchyme contains undifferentiated cells, phagocytic digestive cells, and gland cells. Food is taken in by the mouth and digested intracellularly by the inner digestive cells. Locomotion occurs by means of cilia. In primitive acoels, the nervous system is epidermal and in others there is also a submuscular nerve plexus. An anterior nerve concentration and longitudinal nerve strands are present. A mouth on the ventral surface opens into the mesenchyme either directly or via an intervening pharynx. Intestine, excretory organs, and gonads are lacking. Some very specialized structures are present such as rhabdoids, sagittocysts, and a copulatory apparatus.

According to the first theory, the planuloid form is the common ancestor of the Eumetazoa, giving rise to the Coelenterata and, by way of an acoeloid form, the Bilateria. The second theory derives all the Eumetazoa including the coelenterates from the acoel form. The acoel is somewhat more complicated than the planula, being bilaterally symmetrical and having some specialized features. Both forms have a nervous system, better developed in the acoel.

---

*The epidermis of acoels was formerly thought to be syncytial, but recent observations with the electron microscope have shown that it is composed of separate individual cells (Pedersen, 1964).

4

However, the general body organization and histology of these forms is similar, consisting of an outer layer of cells with locomotor, protective, and food-gathering functions and an inner mass of cells with functions of digestion and reproduction. The existence of a primitive, small, free-living worm-like form with an outer ciliated epidermis and solid interior containing undifferentiated and digestive cells is consistent with both theories. The general organization of this animal may have resembled the ancestor of the Eumetazoa and can be used as a framework in considering the origin of the nervous system. Muscle, gland, and pigment cells were probably present, and nerve cells may have originated in a form similar to this animal. After consideration of some existing primitive nervous systems, an attempt will be made to reconstruct the nature of the primitive nervous system as it could have appeared in the primitive ancestor of the Eumetazoa.

Theories of origin of the nervous system

One of the earliest theories was that of Kleinenberg (1872). He described cells in hydra which he termed neuromuscular cells; they contained an apical region or sensory hair facing the external environment and a basal region drawn out into muscular processes. This single cell type was considered to contain all the components of the nervous arc—the receptor, conductor, and effector. Kleinenberg believed that neuromuscular cells gave rise to nerve and muscle cells containing one of the three components. This hypothesis received little attention, although it was not refuted, when the neuromuscular cell was identified as an epitheliomuscular cell.

The Hertwigs (1878) suggested that the receptor, conductor, and effector arose as separate cell types. All three cells were considered to develop from separate epithelial cells but during this process were physiologically interdependent. Claus (1878) and Chun (1880) suggested that nerve and muscle arose independently, becoming associated only secondarily. The theory of the

5

Hertwigs, in which nerve and muscle develop simultaneously, was generally accepted until Parker's theory appeared.

Sollas (1888) proposed an interesting possibility concerning a nervous system in sponges. Although it does not necessarily represent a theory on the origin of the nervous system, his suggestion indicates some possible features of an early nervous system. He observed in sponges a network of what he considered to be stellate connective tissue cells or collencytes. The processes of these cells connected bundles of myocytes and also extended between pinacocytes and choanocytes. He suggested that the network of collencytes formed a rudimentary type of nervous system.

Since the appearance of his book *The Elementary Nervous System,* in 1919, Parker's views on the origin of the nervous system have become widely accepted and have influenced most subsequent theories. Parker traces, in clear and logical steps, the evolution of the neuromuscular mechanism which includes the sense organs or receptors, the adjustors or central nervous organs, and the effectors such as muscle or glands. Sponges were considered to represent the stage in evolution in which muscle is present in the absence of nerve. In other words, sponges possess effectors but no receptors or adjustors. In coelenterates, receptors make their appearance in the form of a sensory surface or cell. The receptors, sensitive to external stimuli, terminate on the subjacent muscle and thus together constitute a receptor–effector system. The final stage in the development of the neuromuscular mechanism is the appearance of an adjustor or central organ between the receptor and effector. The system of receptor–adjustor–effector is characteristic of more complex invertebrates and the vertebrates.

Sponges represent the first stage in the evolution of the neuromuscular mechanism. The muscular tissue of the pore and oscular sphincters responds directly to environmental stimuli and is an independent effector. Thus, sponges possess effectors in the absence of receptors and adjustors. Other examples of independent effectors are the acontial muscles of sea anemones and the nematocysts of coelenterates. The essential factor in Parker's definition of independent effectors is the capacity of the effector to be stimulated directly. The effector, however, may also be responsive to neural stimuli. For example, the circular muscles of the column of sea anemones may be under nervous control and also be capable of responding directly to environmental stimuli. After feeding, peristaltic constrictions pass down the column of the anemone. Parker considers this activity to be under control of the nerve net. However, if the animal is anesthetized with magnesium sulfate so that contraction of the animal and peristaltic waves are blocked, mechanical stimulation will result in a ring of constriction passing around the column from the point of stimulation. This response is ascribed to direct stimulation of the circular musculature.

Although sponges are considered to be devoid of nerves, Parker points out that they do have some nervous characteristics. For example, when the finger of the sponge *Stylotella* is cut within 1.5 cm of the osculum, the latter closes after several minutes. He believed this slow type of conduction was the elemental property of protoplasmic transmission, and he termed it neuroid transmission. True nervous conduction may have originated from this sluggish type of activity.

The second stage of evolution is the receptor–effector system. Receptors arose by the modification of epithelial cells that were in close proximity to the already differentiated muscle cells. In its simplest form a group of sensory cells would be connected directly to a group of muscle cells. Thus, stimulation of the sensory cells would affect the subjacent muscles. Although there appear to be no examples of this type of arrangement, sensory cells may be situated in the epidermis, with the basal processes forming a network that is connected to the muscle cells. This type of system may occur in the tentacles of sea anemones and represents the beginnings of the nerve net. More commonly in coelenterates, a third type of cell, called a ganglion cell (the Hertwigs, 1879–80), motor cell (Havet, 1901), or protoneuron (Parker, 1918), is interposed between

7

the sensory cells and muscle. This arrangement forms the true nerve net. Although the ganglion cell might appear to be an adjustor, Parker does not believe the nerve net belongs to the final type of neuromuscular organization, because the nervous activities are uncentralized and transmission in the nerve net is diffuse and spreads throughout the animal. Moreover, each portion of the animal is autonomous, possessing its own neuromuscular organization.

The nerve cells comprising the nerve net were considered by Parker to be continuous. The ganglion cell or protoneuron gave rise to the central neuron with the development of synapses in the nerve net. The appearance of synapses was the essential step in the conversion of the nerve net to a true central nervous organ. Synapses polarize the nervous system, allowing transmission to occur in only one of two possible directions. This property is characteristic of the receptor–adjustor–effector system.

In summary, independent effectors, represented by oscular and pore sphincters of sponges, were the first part of the neuromuscular mechanism to arise. Receptors or sensory cells arose next from epithelial cells in close proximity to the effectors. This stage, represented by the coelenterates, constitutes a receptor-effector system. The final step was the appearance of an adjustor or central organ between the receptors and the effectors.

One objection to Parker's theory, raised by Pantin (1956) and Passano (1963), is that there is little evidence for the separate existence of the receptor–effector system composed solely of sensory cells connected with effectors. In coelenterates, some sensory cells may terminate directly on epitheliomuscular or muscular cells, but the majority connect with the underlying network of ganglion cells. Parker did not consider the nerve net to belong to the final stage of organization but to represent a receptor–effector system. However, the nerve net is not a syncytium as he supposed but is composed of individual ganglion cells that are interposed between receptors and effectors. True synapses appear to be present in some coelenterates (Horridge and Mackay, 1962).

Thus, the three elements of the neuromuscular mechanism are already present in the coelenterates.

Parker's theory on the origin and evolution of the nervous system accounted for the anatomical and physiological knowledge available at that time. More recent information including electrophysiological data, fine structural observations, and the occurrence of neurotransmitters and neurosecretory substance have greatly increased our knowledge of the nervous systems of lower animals. In some cases these findings have changed and clarified our concepts of the nature of the primitive nervous system. Although many obvious gaps remain, the present information prompts a reevaluation of the origin and evolution of the nervous system.

Parker's theory was considerably modified by Pantin (1956), who agreed that independent effectors, such as those in sponges, arose prior to nerve cells. Parker's second step was the development of a sensory cell from an epithelial cell and its connection with the effector. Pantin suggested, however, that the behavior of primitive animals is not the result of the individual contractions of single cells but is due instead to the coordinated contraction of large regions or fields of muscle sheets. In other words, the behavior machine did not evolve by the stepwise addition of simple reflex arcs, because the response of an individual cell to a stimulus would be insignificant in relation to the entire muscle sheet. Pantin believes that the contractile system, operating as a whole, has a considerable range of activities in the absence of a nervous system. The nerve net was superimposed on the integrated muscle sheet and increased the area of the muscle sheet responding to a single stimulus. The nerve net may have provided a pathway for conducting excitations that was different from the conducting pathway in the contractile sheet. Furthermore, specific conducting tracts associated with specific reflexes appeared in the nerve net and gave rise to the reflex arc which, by this view, is not primitive.

Passano (1963) proposed a modification of Pantin's theory in which individual protomyocytes evolved into groups of contractile cells, some of which became centers of endogenous activity, or pacemakers. The endogenous activity of pacemakers affected the contractile mechanisms. Depolarization may have spread over specialized intercellular bridges. Muscle cells and pacemakers then evolved simultaneously, the latter becoming specialized for conduction of activity. Pacemaker cells would have been influenced by external stimuli and later, as nerve cells (sensory cells), integrated their recurring internal activities with rhythmic external stimuli. Subsequent evolution of the nervous system involved the appearance of through-conducting tracts associated with specific reflexes, the concentration of nerve cells into ganglia or nerve rings, and the accumulation of sensory cells in organs associated with the ganglia.

Pantin (1965) has pointed out that it may be difficult to construct a model in which pacemakers are correlated with complex behavior. Nevertheless, pacemakers have been correlated with some of the activities performed by hydra (McCullough, 1965) and could have a relationship to the behavior of a simpler animal. Although it would be surprising if nerve and muscle have a common origin as Passano implies, Pavans de Ceccatty (1966a) described cells in sponges with both neural and muscular specializations.

Grundfest (1959, 1965) has suggested that nerve cells arose from ancestral secretory cells. The sensory cell was developed by the specialization of receptive surfaces with sensitivity to specific stimuli so that this cell had receptor and secretory poles. The receptive and secretory portions of the cell gradually were displaced but remained connected by a region with conductile properties. The development of long processes terminating near blood vessels or on other cell types led to the differentiation of neurosecretory cells. Neurons were formed when the secretory activity became confined to the terminations of the processes. This theory is of particular interest because it takes into consideration current physiological and morphological evidence and explains the origin of the different nerve cell types. Clark (1956a,b) similarly derives neurosecretory cells from secretory epidermal cells: in conventional neurons, the basic primitive secretory ability has become more specialized and restricted.

This brief review of some of the important theories on the origin of the nervous system outlines the trends in thought that have taken place. The earlier theories were based largely on anatomical descriptions. More recently, behavioral and electrophysiological findings have been taken into greater consideration. Finally, the importance of neurosecretory cells has been recognized, and the possibility that secretion is a primitive feature of the nervous system is being raised. It should be apparent that a theory based on the findings of one field alone is likely to be incomplete. It is necessary to correlate and account for the findings from all areas of investigation. I do not intend to review the merits of each theory here. Most of the theories emphasize certain points or raise important questions, and these, as well as some other suggestions which have been made, will be considered as they arise in the following chapters.

Definition of a nervous system

Any theory on its origin must begin with a clear definition of a nervous system, to establish the characteristics that must be present before a cell can be considered a neuron. Although this may seem apparent, confusion has arisen from lack of agreement concerning a definition or from failure to establish one. To some extent the definition may be arbitrary and depend on how many features of a differentiated system are postulated as requirements before a primitive or simple nervous system can be said to exist. For this reason, a somewhat broad interpretation may be preferable. The definition that will be followed here is that of Bullock and Horridge (1965b), who define a nervous system as

> an organized constellation of cells (neurons) specialized for the repeated conduction of an excited state from receptor sites or from other neurons to effectors or to other neurons.

A neuron, then, is a cell specialized for the reception of stimuli, conduction of excitation, and transmission of the signal to other cells.

Nerve cells in widely separated groups employ similar mechanisms to carry out the basic requirements. It might be helpful if I summarize the functions of a generalized neuron, recognizing that there is some variability and that some of the processes are poorly understood. Cell membranes are polarized with the interior negative relative to the exterior. Excitable cells are capable of changing the permeability of the cell membrane to specific ions in response to external stimuli. The change in ion permeability results in depolarization or decrease in the resting potential of the membrane. The cell then undergoes a reaction, known as excitation, which enlarges the initial small depolarization caused by the stimulus. This response can be graded or all-or-none. The graded response is an excitation whose intensity depends on the strength of the stimulus; it is local and conducted decrementally. If the initial depolarization reaches a threshold level, an all-or-none response or spike is produced. The spike excites other regions of the membrane and is thus propagated without decrement. In the nerve ending the electrical message usually is coupled to the release of neurohumors or transmitters. These agents diffuse across intercellular spaces and affect the permeability properties of the membranes of other neurons and effectors, leading to a response by these cells. Electrical, electrotonic, or ephaptic transmission is accomplished by direct current flow from the pre- to postsynaptic cell. Although the primitive nerve cell was probably not so specialized, this scheme will be useful later in discussing neural functions.

Welsh (1955) has pointed out that the most characteristic physiological feature of the nerve cell is the production and release of biologically active chemical agents that integrate bodily functions. The chemical agents are utilized as messengers in communication between cells, whereas electrical phenomena are more prominent in the spread of excitation within a cell.

Pantin (1952) has outlined three criteria for the identification of nerve cells in coelenterates, which also appear applicable to other lower groups: the cells should stain in the same manner as nerves in other animals; their cytological structure should resemble that of other nerve cells; and their anatomical relationships should be consistent with the physiology of the animal. The first two points should not be interpreted as meaning that primitive neurons are the same as evolved neurons. They can be expected, however, to possess some, and probably the basic and most primitive, features of higher neurons.

Identification of neurons in lower animals is even more difficult on the fine structural level and usually demands that the cells observed in thin sections correspond to those called neurons under light microscopy. In some cases it is possible to identify structures (microtubules, vesicles, membrane-bounded granules, synapses) characteristic of higher neurons. The difficulty remains that the structure of primitive neurons may not be the same as that of higher neurons or that the cells thought to be neurons in histological preparations are actually some other cell type. These problems emphasize the need for more observations, based on as many of the criteria that can be applied, on a variety of lower invertebrates.

2

THE NEURAL ORGANIZATION
OF SPONGES

Introduction

The Porifera (sponges) are the lowest group of metazoans and, because of their unusual characteristics, are separated from the rest of the Metazoa and placed in a branch of the animal kingdom named Parazoa. All the other Metazoa comprise the branch Eumetazoa, or true Metazoa. Sponges are considered to represent a cellular grade of construction in which the organism is composed of a loose aggregation of cells that are not organized into specialized tissues.

The simplest and smallest sponges are vase-shaped but the majority are without definite shape and assume fan-like, branching, or encrusting forms. They are sessile and are fastened to mud, rocks, or shells. The body of sponges is porous and permeated with channels for the circulation of water which enters through numerous incurrent ostia over the surface of the animal, circulates through interior channels, and leaves through larger and less numerous excurrent openings or oscula. The simpler sponges contain a large central cavity or spongocoel.

At the histological level (Fig. 1), sponges are composed of an inner and outer epithelium and an intervening mesenchyme. The outer epithelium or epidermis is composed of a single layer of thin flat cells or pinacocytes. In some sponges, pore cells or porocytes occur in the epidermis around pores. Most of the inner epithelium is formed by choanocytes or collar cells. The flagellar activity of choanocytes is responsible for the circulation of water. The mesenchyme is composed of a gelatinous matrix within which are embedded spicules and several types of cells. There are different types of wandering amoeboid cells or amoebocytes including scleroblasts which secrete the skeleton, food-storing cells or thesocytes, pigmented chromocytes, and collencytes. Archeocytes are undifferentiated embryonic cells. Myocytes are contractile and occur around the osculum or other openings. Sponge myocytes contain both thick and thin filaments and closely resemble smooth muscle cells of other invertebrates (Bagby, 1966). Other mesenchymal cells are gland cells, desmacytes or fiber cells, and, according to some workers, nerve cells. The skeleton occurs in the mesenchyme and consists of spicules and spongin fibers.

Most investigators have concluded that sponges contain no nerves or sensory cells; the main argument against their presence is that the simple behavior of these animals does not require a specialized conducting system. Those activities of which sponges are capable are considered to be due to the action of independent effectors that respond directly to external stimuli.

The question of the presence of a nervous system in sponges deserves additional study, for several reasons. First, if the behavior of an animal is simple and might seem not to require a nervous system, it does not necessarily follow that a nervous system is absent. A primitive nervous system, especially, might be associated with simple behavior. In addition, the concept of independent effectors does not exclude the presence of nerves, because neural influences may alter the activity of cells that respond directly to external stimuli. Finally, if sponges are totally devoid of any neural specializations, there is no neural level of organization between the Protozoa, which lack a nervous system, and the coelenterates. The latter represent the lowest metazoans in which a nervous system is definitely known to exist. The nervous system of coelenterates, however, is complex and specialized in many respects. It would appear, therefore, that the search for a more primitive existing nervous system or an organization approaching a nervous system should be undertaken in the Porifera, which exhibit the lowest grade of metazoan organization.

Although some evidence indicates the absence of a nervous

Fig. 1. Histology of the body wall of *Sycon*. Pinacocytes (P) form the outer epidermis and line the incurrent canals (InC) and the spongocoel (Sp). The radial canals (RC) are lined by the flagellated choanocytes (Ch). The mesenchyme occupies the space between the inner and outer epithelia. The cell types in the mesenchyme include amoebocytes of several types such as archeocytes (Ar), scleroblasts (Scl) associated with spicules, and cells filled with vacuoles. Myocytes or muscle cells (MC) encircle a prosopyle (Pr). Cells observed in histological sections thought to correspond with those staining histochemically for neurohumors are bipolar (Bp) and multipolar (Mp) cells. The bipolar cells occur beneath the pinacocytes; the larger multipolar cells occur throughout the mesenchyme. The arrow traces the path of water circulation. O, ostium; IO, internal ostium.

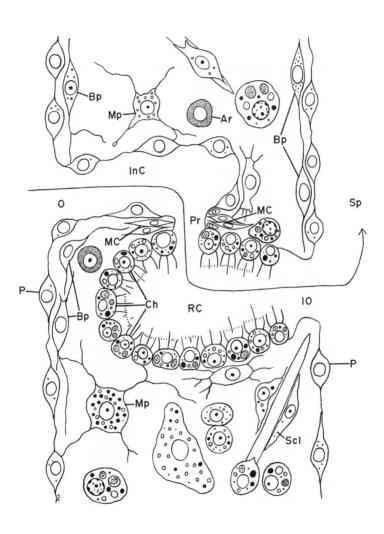

system (see Jones, 1962), other findings suggest that nerve cells indeed do exist in sponges. Much of the controversy and confusion regarding this conflicting evidence might result because the nervous system, if it does exist in these animals, is probably very simple and might not exhibit all the features of an evolved system. Any nervous system, of course, must fulfill the criteria established in the definition. Even if the minimum requirements are not met, it is possible that certain cells in sponges have some neural characteristics. In this case the level of organization exhibited by sponges may reflect the nature of the earliest nervous system or a stage just prior to the origin of the nervous system.

Physiology

Adult sponges remain attached to the substrate and are not capable of locomotion, although amoeboid movement may occur in very young animals (Metschnikoff, 1879). Sponges are capable of several simple activities and responses. Body contraction can occur in response to a variety of stimuli including mechanical stimulation, injury, quiet water, exposure to air, change in temperature, and harmful chemicals. Contraction may also occur in response to light (Wintermann, 1951) and electrical stimulation (McNair, 1923; Pavans de Ceccatty et al., 1960). These contractions can be localized or generalized, with the entire animal rounding up. The alterations in shape occur as a result of contraction of pinacocytes, desmacytes, and myocytes. The color of sponges is also affected by light (Orton, 1924).

Water is circulated by the flagellar action of choanocytes, entering through dermal ostia, circulating through the sponge, and flowing out the osculum. Food is carried to the interior on water currents and is phagocytosed by the choanocytes where it is digested or more usually passed on to the amoebocytes in the mesenchyme. The rate of flow of water appears to be regulated by the degree of expansion of the osculum and pores, rather than by alteration of flagellar activity (Parker, 1910). Normally the oscula and dermal pores are open. Closure of the osculum by a ring or sphincter of myocytes occurs after the same stimuli that elicit body contraction, including light (Jones, 1957). Dermal pores also undergo closure, usually after harsh mechanical stimulation or injury. All the movements performed by sponges are very slow, often taking minutes to complete.

Sponges show limited conductivity. McNair (1923) described a contraction wave passing down the chimney after mechanical stimulation of the osculum; it traveled 1.5 mm in 6 seconds (0.25 mm/sec). After electrical stimulation, it moved 1.5 mm in 4 seconds (0.375 mm/sec). The greatest reported distance of conduction occurs in *Tethya,* in which stimulation of the base results in oscular closure 4 to 8 cm away (Pavans de Ceccatty et al., 1960). Attempts to record electrical potentials in sponges have been unsuccessful (Prosser et al., 1962).

Parker (1910) tested the effects of some chemical agents on dermal pores and oscula of *Stylotella (Hymeniacidon).* Ether, chloroform, strychnine, and cocaine produced closure of pores and oscula. Atropine and a weak solution of cocaine caused opening of the pores but inhibited closure of the oscula. Parker considered that these responses resemble those of vertebrate smooth muscle. Jones (1957) found that magnesium chloride had no affect on the contractile behavior of sponges, and Prosser et al. (1962) observed that mechanical responses persist in the presence of potassium.

The effects of a number of pharmacological agents on *Cliona* were tested by Emson (1966). The agents studied included adrenalin, acetylcholine, cocaine, hexamethonium, glutathione, eserine, adenosine triphosphate, serotonin, γ-amino butyric acid, nicotine, histamine, atropine, D-tubocurarine, and tryptamine. Closure of the osculum occurred in some cases after exposure to drugs but was considered to be the result of a toxic action on the choanocytes. Studies of this type, especially the effects of neurohumors and their inhibitors, are of great importance in determining the nature of the neural organization of sponges. Of particular interest would be studies of the effects of these agents on response to other types of stimulation, such as light or mechanical, because the neurohumors may function to modulate or alter behavior rather than to initiate it.

Histology

One of the earliest descriptions of nerve cells in sponges was a diagram by Stewart (1885) in a textbook, showing sensory cells in *Grantia* around the external orifices of the interradial canals. These cells have distal processes and it was suggested they might be related to regulation of water currents. Lendenfeld (1885a,b,c, 1886, 1889, 1892) made more extensive observations on presumed nerve cells in sponges. He described cells that appear similar, when stained with osmic acid, to the ganglion and sensory cells of coelenterates. Two types of cells called sensitive cells are near the epidermal surface; the first is elongated and spindle-shaped with a distal unbranched process and a basal branching process; the second is multipolar with a single long distal process and numerous short basal processes. These cells, or at least the distal processes, are perpendicular to the surface of the sponge. Lendenfeld also described ganglion cells which are multipolar and situated beneath the sensitive or sensory cells. The basal processes of the sensory cells often anastomose with processes of the ganglion cells. Lendenfeld later (1892) described sensory organs in *Sycon* termed synocils; these are groups of sensory cells in conical elevations on the surface of the dermal epithelium.

More recently, Tuzet and Pavans de Ceccatty have described nervous or nervous type cells in a large number of species by the use of silver impregnation and other techniques (Tuzet et al., 1952; Tuzet and Pavans de Ceccatty, 1953a,b,c,e; Pavans de Ceccatty, 1955; see also Jones, 1962, for review). The main types of neurons which they described are classical neurons, vesiculous neurons, spider cells, and sensory cells. Neurons of classical or standard *(classique)* type were identified in a large number of genera including *Sycon, Grantia, Leucandra, Clathrina, Cliona, Halichondria, Pachymatisma,* and *Halisarca.* These cells are fusiform or triangular in shape, and some are multipolar. They contain a large central nucleus but the nucleo–cytoplasmic ratio is small. The cytoplasm contains silver-staining particles thought to represent Nissl substance. Intracellular fibrils, possibly representing neurofilaments, were observed. Processes extend from the cell body, and two types have been distinguished. Numerous fine short processes, corresponding to dendrites, arise from one pole of the cell, and a single axon-like process extends from the other side. The processes contain varicose enlargements along their length; they connect neurons with pinacocytes, choanocytes, myocytes in the walls of canals, supporting fibers in the mesenchyme, and other neurons, or they may terminate in intercellular spaces. Arcs of more than one neuron may occur between these structures. The distal processes form terminal arborizations. Nerve endings on cells other than neurons consist of intracellular boutons projecting into the innervated cells. Continuity between neurons is by protoplasmic fusion.

Vesiculous or bladder *(vésiculeux)* cells were found in *Halichondria* and *Pachymatisma*. These cells have a reticular or vesiculous cytoplasm and are thought to be derived from the pinacoderm. They are initially amoeboid but later form a network and may be capable of giving rise to other nerve cell types.

Spider or arachnoid *(araigneé)* cells have been identified in *Leucandra, Cliona, Pachymatisma, Leucosolenia,* and *Halichondria.* These large cells are usually inside vacuoles or spaces in the mesenchyme. Several fine processes and a single thick axon arise from the cell body. The processes may terminate within the vacuole, in its wall, or extend into the mesenchyme connecting with neurons or other cell types. Somewhat similar cells in *Grantia* were considered aberrant preganglionic neurons.

Neuromuscular cells, thought to have conducting and contractile functions, were identified in *Cliona* and *Halichondria.* These cells are spindle-shaped with a single axonal process at one end and a large number of thinner fibers radiating from the other end. The axon joins the network of classical-type neurons; the fine processes envelop a spicule bundle or the wall of a canal. These cells often extend across cavities or spaces.

Sensory cells were described in *Leucandra* and *Hippospongia* and very small numbers were observed in *Sycon*. Most sensory cells are spindle-shaped and are perpendicular to the surface. A distal process may project above the epithelium. The proximal process may connect with neurons of the classical type or extend to the layer of choanocytes. Peculiar cells known as lophocytes and possessing a tuft of fibers are not considered nervous elements (Ankel and Wintermann-Kilian, 1952; Tuzet and Pavans de Ceccatty, 1953d).

Neurochemistry

Histochemical techniques have revealed neurohumors and related enzymes in some cells of *Sycon* (Lentz, 1966b). The substances identified were acetylcholinesterase, monoamine oxidase, epinephrine, norepinephrine, 5-hydroxytryptamine, and neurosecretory substance. Two types of cells contain the most intense staining for these substances. Spindle-shaped cells occur in the mesenchyme, closely applied to the inner side of the epidermal pinacocytes or the layer of choanocytes (Figs. 1, 2). These cells are usually bipolar with long processes. They are very numerous, oriented circularly in the inner aspect of the rim of the osculum and in the membrane extending across the osculum. The processes of the circularly oriented cells in the osculum overlap. In the neck or collar of the sponge between the osculum and the first canals, the cells are oriented approximately in the long axis of the sponge. In this region the processes have a few branches and interconnect to form a network. Most of the cells are arranged into longitudinal strands. There are a few spindle-shaped cells between the canals, and they are also numerous near the base of the sponge.

The second reactive cell type is a large multipolar cell with three to ten short processes (Figs. 1, 2) which is present in all regions and is most abundant in the collar below the osculum. Sometimes the processes of adjacent cells are in contact, but most appear to terminate in the intercellular spaces. The multipolar cells are also situated in the mesenchyme but appear to bear no close relation to epidermal cells. Cells that might correspond to sensory cells were observed rarely with the histochemical techniques.

Small granules, vesicles, or larger droplets are reactive, depending on the method employed. Stained vesicles are often present in the processes and sometimes are accumulated in bulbous swellings along the processes. Acetylcholinesterase activity occurs in both the spindle-shaped and multipolar cells and is localized to small cytoplasmic granules. Monoamine oxidase activity also occurs in both cells, but especially in the multipolar cells where it is localized to small vesicles and droplets in the cytoplasm and processes. Epinephrine stains more prominently in the multipolar cells (Fig. 2). The reaction occurs in large cytoplasmic droplets. Few stained droplets occur in the processes. Norepinephrine, on the other hand, stains predominantly in the bipolar cells (Fig. 2). Staining appears in vesicles or granules smaller than the droplets that stain for epinephrine. Norepinephrine also stains more prominently in the processes. 5-Hydroxytryptamine is more apparent in multipolar cells and occurs in large vacuoles and small granules. The activities or staining of these substances could be blocked by the use of appropriate inhibitors.

Neurosecretory substance was revealed by Bargmann's chrome hematoxylin method and Alcian blue staining after oxidation. Neurosecretory substance as revealed by these methods occurs in droplets within the bipolar and multipolar cells. Bulbous swellings containing stained vesicles are present in the processes. Unlike the other reactions, some cells show diffuse cytoplasmic staining. Chrome hematoxylin staining is most intense in the multipolar cells, whereas Alcian blue is more prominent in the bipolar cells. The network formed by the processes of bipolar cells was especially well demonstrated by Alcian blue. It is recognized that these methods may stain substances other than neurosecretory and provide only presumptive evidence for this type of cell. Morpho-

Fig. 2. Cells in the mesenchyme of the sponge *Sycon* as revealed by epinephrine and norepinephrine staining. Two cell types stain for these substances as well as with the other histochemical methods employed. One cell type is spindle-shaped and overlies the second type. These bipolar cells form longitudinal strands and show a more intense staining reaction for norepinephrine. Staining occurs in small vesicles or droplets in the cytoplasm and along the processes. The second cell type is large, with many processes, some of which seem to terminate in bulbous enlargements; but most could not be traced for any distance. The multipolar cells show a more intense localization for epinephrine. Staining occurs in large vacuoles and droplets.

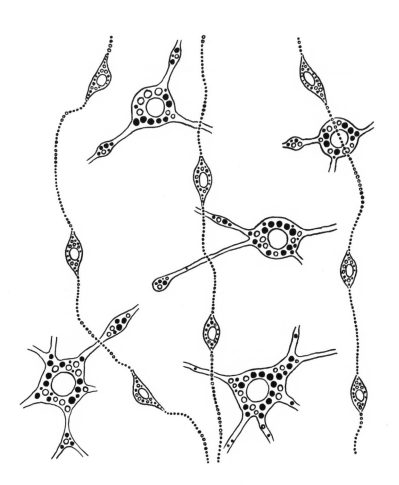

logical evidence for neurosecretion is also available (see following section) but physiological criteria have yet to be fulfilled.

Neurohumors have not been identified in sponges by biochemical or assay procedures. Studies have been performed on acetylcholine (Bacq, 1935, 1947), acetylcholinesterase (Mitropolitanskaya, 1941; Bullock and Nachmansohn, 1942), and 5-hydroxytryptamine (Welsh and Moorhead, 1960). The reason for the discrepancy between the biochemical and histochemical studies is not known. It is possible that there are species differences or that adequate tissue samples were not available for some of the biochemical determinations. Florey (1962) pointed out that in studies of the acetylcholine content of sponges and coelenterates, the amount of nerve tissue that could be present in the material used for bioassay is negligibly small, and that the apparent absence of this substance does not necessarily argue against its presence in the nervous system. It is also possible that the histochemical reactions are nonspecific and demonstrate substances related to those for which the techniques were designed. For example, unidentified reducing substances might give a positive chromaffin or argentaffin reaction. A substance related to catecholamines was thought to occur in sea anemones (Wood and Lentz, 1964). Östlund (1954) has suggested that a substance behaving like a catecholamine might be biologically active in some invertebrates. Although the methods might be nonspecific, there is a reasonably good correlation between histochemical data and other methods for the identification of neurohumors in coelenterates, planaria, and other invertebrates.

The simple responses of which sponges are capable are usually thought to be due to the activity of independent effectors such as the myocytes of the osculum. However, if these effectors are sensitive to biologically active neurohumors, the cells in sponges containing these substances might be able to affect or modify the activity of the effectors. The chemically specialized cells are loosely arranged and often appear to terminate in intercellular spaces. Whatever control these cells might have on effectors is probably diffuse, slow acting, and not effective over great distances. It is not known, however, whether neurohumors can alter the activity of sponge effectors or are released in response to external stimuli. In the absence of this evidence, these cells cannot be regarded as nerve cells. It is more reasonable, at the present time, to consider them a recognized sponge cell type (e.g. amoebocyte, collencyte) that has some neural specializations. Because of these specializations, they may represent a primitive situation preceding the appearance of neurons.

Fine structure

The histochemical studies indicate that some cells in sponges are specialized for the production of neurohumors and neurosecretory substance. In higher animals some of these substances (e.g. epinephrine, norepinephrine, and neurosecretory substance) are contained in well-defined structures (membrane-bounded granules) that show relatively little difference in widely separated groups of animals. It can be expected, therefore, that similar structures might occur in those sponge cells containing neurohumors. In order to investigate this possibility, *Sycon* was studied under the electron microscope. It is possible to observe a few cells in thin sections of undecalcified sponges, although most are obscured by the numerous scratches and gouges produced by the spicules during sectioning. Spicules can be removed by treatment for three days (4°C) in 0.05 M ethylenediamine tetraacetate (EDTA, Versene) after glutaraldehyde fixation and before osmium tetroxide postfixation. Considerably better sections can be obtained with this method, although the morphology is not so well preserved.

Two cell types were observed in *Sycon* with the electron microscope that contain features present in more evolved neurons (Figs. 3, 4). In most respects these cells are similar. Both are located in the mesenchyme and are sometimes near the basal surface of pinacocytes or choanocytes. The cells are either irregular in shape with several cytoplasmic processes or are oval,

27

Fig. 3. Granule-containing cell in the mesenchyme of the sponge *Sycon;* it is large, irregular in shape, and has cytoplasmic extensions or processes. The central nucleus contains a large nucleolus. The cell contains a large number of cytoplasmic dense granules (DG). The granules are of medium density, bounded by a membrane, and are 1,100 to 1,700 A in diameter. They are distributed throughout the cytoplasm and in some cells, like the one illustrated, they are very abundant. A large Golgi apparatus (G) consists of membranous lamellae, vesicles, and membranous sacs. Dilated cisternae of rough-surfaced endoplasmic reticulum, mitochondria, and multivesicular bodies (MVB) occur in the cytoplasm. Large vacuoles (Va) containing dense particulate material resemble the large phagocytic vacuoles of amoebocytes. Peculiar curved, crescent, and doughnut-shaped bodies are found in most sponge cells. C, centriole.

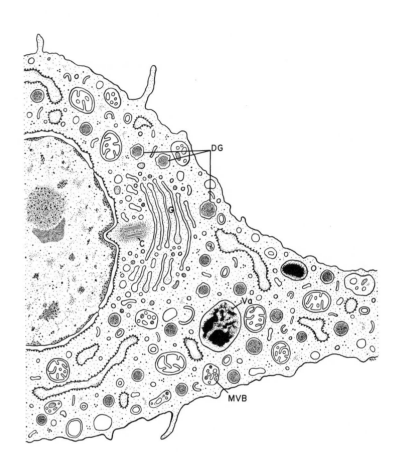

spindle-shaped cells. They contain a large, round, central nucleus that is bounded by an envelope containing pores. The nucleoplasm is of relatively low density except for the chromatin material that is condensed peripherally. A large eccentrically located nucleolus is composed of fibrous material and dense granules, 150 A in diameter. The hyaloplasm is of low density and contains free ribosomes. Mitochondria are small, oval or round, with few cristae. Elements of the endoplasmic reticulum are sparse, consisting of a few round or irregular cisternae, some containing ribosomes on their outer surface. Membrane-bounded structures, 0.2 to 0.4 μ in diameter, are filled with small vesicles and resemble multivesicular bodies. Larger vacuoles (up to 0.5 μ) contain dense granular or amorphous material. These structures resemble phagocytic vacuoles. The cells contain a prominent Golgi apparatus. A centriole sometimes occurs between the Golgi zone and nucleus.

The two cells are distinguished by their content of cytoplasmic dense granules which are surrounded by a membrane and are 1,000 to 1,700 A in diameter (most ~ 1,400 A) with moderately dense contents (Fig. 3) or ~800 A in diameter with electron-opaque contents (Fig. 4). The granules occur in the perikaryon where they are sometimes very numerous and in the processes of these cells. They occur in close relation to the Golgi apparatus. This organelle is adjacent to the nucleus and consists of a stack of membranous lamellae and small vesicles. The vesicles are about 500 A in diameter and contain material of low or medium density. Dense granules are abundant in the Golgi region, especially near the ends of the lamellae.

A few isolated processes containing granules identical with those in the cell bodies were observed. Nothing resembling a synapse or even a nerve ending was noted, although granule-containing cells or processes may be situated adjacent to other cell types.

Because little information is available on the fine structure of sponges, some mention should be made of the other mesenchymal cells to distinguish them from the granule-containing cells. A few cells are small, round, and filled with free ribosomes; other organelles are few in number. These cells resemble hydra interstitial cells and planarian neoblasts and are presumably undifferentiated (archeocytes). Some cells contain oval, electron-opaque granules about 0.3 μ in length and resemble the pigment cells of higher animals. Others, probably scleroblasts, are closely associated with the spicules or small extracellular fibrils and contain numerous dilated cisternae of endoplasmic reticulum. The majority of the mesenchymal cells are irregular in shape with blunt or elongate processes and are filled with large (1–3 μ) vacuoles containing dense particulate material. These cells are clearly phagocytic and correspond to amoebocytes. They also contain a curious structure which is a curved crescent-shaped space bounded by a membrane. It looks like a portion of hyaloplasm indenting or invaginating into a small vacuole. These structures are present in most other sponge cells including pinacocytes, choanocytes, and granule-containing cells. Some mesenchymal cells contain large membrane-bounded granules (0.5–2 μ) that are electron opaque or moderately dense. These structures resemble mucous granules.

Granule-containing cells can be distinguished from amoebocytes, although there are some similarities in the two cells. Some amoebocytes may contain one or two small dense granules but most are devoid of them. Amoebocytes are filled usually with large phagocytic vacuoles. Similar vacuoles have been observed in a few granule-containing cells, but these are not numerous and do not exceed 0.2 μ in diameter. The Golgi apparatus is not so complex in amoebocytes and does not have granules associated with it. Finally, the crescent-shaped structures common in amoebocytes and other cell types are less numerous in granule-containing cells.

Pavans de Ceccatty (1966a) has described in the sponge *Tethya lyncurium* the fine structure of mesenchymal cells, which he previously identified as nervous by silver impregnation methods (Pavans de Ceccatty, 1960). The cells are irregular in shape, with processes, and have a central nucleus containing a nucleolus.

Fig. 4. Second type of granule-containing cell observed in the mesenchyme of *Sycon*. This cell is smaller and more regular in shape. The central nucleus contains a nucleolus. The granules (DG) of this cell are opaque and about 800 A in diameter. All are bounded by a membrane. The granules are most abundant at the periphery of the cell. A well-developed Golgi apparatus (G) with elongated lamellae occurs near the nucleus. The lamellae and vesicles of the Golgi contain moderately dense material. Cisternae of rough-surfaced endoplasmic reticulum (ER) are not numerous but are dilated and contain fine particulate material. A few free ribosomes occur in the cytoplasm. Mitochondria and crescent-shaped bodies are also present.

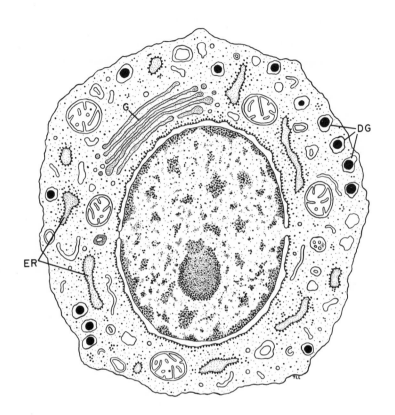

The cytoplasm contains mitochondria, a poorly developed endoplasmic reticulum, and several types of inclusions. A Golgi apparatus is usually present. In the cytoplasm are dense membrane-bounded granules which may occur near the Golgi apparatus or at the level of the processes; they are most often about 1,500 A in diameter. Pavans de Ceccatty noted their similarity to the neurosecretory granules of higher animals. There are microtubules in these cells as well as two types of filaments, the first of which is about 60 A in diameter and often occurs in bundles, whereas the second type is 80 to 130 A in diameter and is more randomly distributed. The two types of filaments may occur in the same cell but are not associated with each other. The relationship of the filaments of sponge cells to known types of filaments (myofilaments, tonofilaments, gliofilaments, neurofilaments) was discussed by Pavans de Ceccatty. The possibility that these structures are myofilaments raises the question of the simultaneous occurrence of muscular and nervous specializations in the same mesenchymal cells.

Several types of connections were observed between cells of the mesenchymal network. Protoplasmic continuity of processes of different cells was thought to occur. Other junctions consist of appositions of adjacent plasma membranes. Simple parallel junctions are separated by an intercellular space of 100 to 150 A. Sometimes an accumulation of dense granules occurs in the cytoplasm of one cell adjacent to the junction. In other junctions, separated by 100 to 125 A, there is a local differentiation consisting of dense material on the inner sides of the opposing membranes. The intercellular substance appears to bear some organization. The last type of junction occurs between mesenchymal cells and flagellated endopinacocytes. These endings consist of terminal swellings or boutons projecting into indentations on the surface of the pinacocytes. Filaments and vesicles are found in the boutons. The intercellular space of these junctions is 200 A. Some boutons were also observed on scleroblasts. This type of junction was thought to function both for mechanical anchorage and transmission between cells. The possibility was raised that the pinacocytes might have a sensory function.

Pavans de Ceccatty (1966b) also studied intercellular relationships in *Hippospongia* at the fine structural level. Desmosomes with typical intracytoplasmic filaments associated with them were observed between some cells (e.g. pinacocytes). Another type of junction consists of a cytoplasmic extension protruding into an indentation on the surface of another cell. The knob-like protrusions contain dense irregular granules, 200 to 500 A in diameter, and vesicles, 600 to 1,500 A in diameter. According to Pavans de Ceccatty, these polarized junctions sufficiently resemble synapses to suggest functional relationships between cells.

The fine structural observations support the indications of the other morphological and the histochemical data that some cells in sponges have some of the characteristics of neurons. However, until a relationship between the activity of these cells and the behavior of sponges is established, they should not be regarded as nerve cells. On the basis of the present information, therefore, I consider that these cells have neural characteristics but do not satisfy all the criteria for a true nervous system. As discussed in the final chapter, this situation may be very primitive and similar to the condition that shortly preceded the appearance of the nervous system.

3

THE NERVOUS SYSTEM
OF HYDRA

Introduction

The coelenterata are the lowest Eumetazoa. They contain specialized tissues and represent the tissue grade of construction in contrast to the cellular construction of sponges. There are three classes
in this group: the Hydrozoa; Scyphozoa, which includes the
jellyfish; and the Anthozoa, containing the sea anemones. Most
coelenterates are marine. They may take the form of a polyp or a
medusa and consist of an outer epidermis and inner gastrodermis
enclosing a hollow digestive cavity (Fig. 5). The two cell layers
are separated by the mesoglea; the epidermis is composed of
epithelial cells which occur in several forms. In hydroid polyps
and parts of hydroid medusae, the epithelial cells, known as
epitheliomuscular cells, contain a drawn-out base containing myofilaments. In other coelenterates, separate muscle cells occur at the
bases of the epithelial cells. Some epithelial cells are specialized
for the secretion of mucus. A characteristic of the phylum is the
presence of nematocysts which are complex organoids contained
within cnidoblast cells. Interstitial cells are undifferentiated and
are capable of differentiating into other cell types.

Networks of neurons occur at the bases of epithelial cells
(Fig. 5). Sensory cells are also numerous, especially in the epidermis. The gastrodermis is composed of columnar nutritive or digestive cells and gland cells which secrete mucus and digestive enzymes. The mesoglea of hydrozoan polyps is a thin acellular layer
but in medusae it comprises the bulk of the animal. In the scyphomedusae, the mesoglea contains amoeboid cells. In the Anthozoa,

the mesoglea forms a mesenchyme containing cells and numerous fibers.

In many respects the nervous system of coelenterates shows a high degree of structural and functional specialization. Many characteristics of more evolved nervous systems are displayed in this phylum. The nervous system reaches its highest level of organization in the free-swimming medusae whose neurons are concentrated into marginal ganglia and nerve rings. In the scyphozoan medusae, sensory organs such as ocelli, statocysts, and patches of sensory epithelium (olfactory pits) occur. In sedentary polyps, the nervous system is not so highly developed and consists of isolated epidermal ganglion and sensory cells. The anatomical and functional gap between the level of organization in sponges and the medusae is much greater than that between sponges and coelenterate polyps. Because the nervous system of polyps appears to represent an intermediate stage of structural and functional complexity and organization between the beginnings of the nervous system and the appearance of a central nervous system in flatworms, it is considered in detail here, with the fresh water hydra as a typical example of this level of organization.

Physiology

Hydra are capable of a number of simple activities and responses. Resting, undisturbed hydra are typically attached to the substrate by the base or float on the surface of the water with the base upward. The base is capable of secreting a bubble of gas that carries the animal to the surface. In resting position the body is extended; the tentacles are elongated and hang down, drifting with the water currents. Hydra periodically contract into a tight ball as a result of contraction of the longitudinal musculature. Partial contraction of the musculature results in bending, twisting, swaying slowly from side to side, or waving of the tentacles.

Movement from place to place occurs in several ways. Trembley (1744) described the now familiar somersaulting movement in which the base and head attach to the substrate alternately. One

Fig. 5. Histology of hydra and its nervous system. The animal is composed of an outer epidermis and inner gastrodermis. The two cell layers are separated by the mesoglea (Me). Ganglion cells (GC) are spindle-shaped cells at the bases of the epitheliomuscular (EmC) or digestive (DC) cells. Sensory cells (SC) are perpendicular to the surface and have an apical specialization. Their basal fibers join the plexus at the bases of epithelial cells. Neurosensory cells (NsC) are ganglion cells with a process extending to the surface. The epidermal system of cells is more extensive than the inner gastrodermal, and fibers do not extend across the mesoglea. The nerve fibers may have swellings along their length and terminate on the muscular processes (MP) of the epithelial cells, on cnidoblasts (Cd), or in the intercellular spaces. GlC, gland cell; IC, interstitial cell.

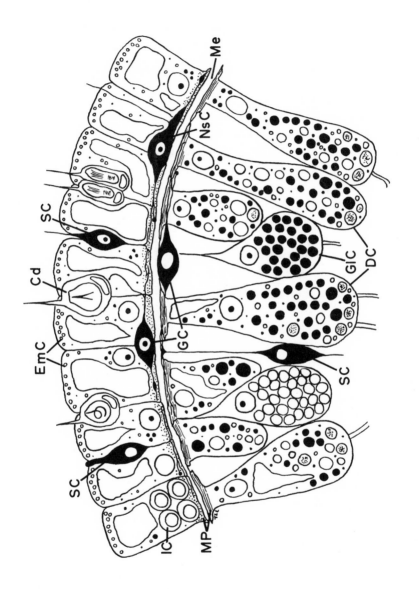

type of nematocyst, the atrichous isorhizas, may fasten the tentacles to the substrate during locomotion (Ewer, 1947). Movement also occurs by basal gliding and attachment of the tentacles to the substrate followed by a contraction of the body which drags the animal along.

Hydra respond to external stimuli such as light, temperature change, injury, and mechanical, chemical, and electrical stimulation. They respond to light by moving toward it (Wilson, 1891), contracting (Haug, 1933; Rushforth et al., 1963; Singer et al., 1963), or ceasing periodic contractions of the body (Passano and McCullough, 1962). Hydra are most sensitive to light in the blue range; receptors to mechanical stimulation appear to act independently of light receptors (Rushforth et al., 1963).

Rushforth (1965) found that removal of the tentacles of hydra blocks the contraction of the animal in response to mechanical agitation. He suggested that sensory cells, numerous at the bases of the tentacles and sending processes to ganglion cells, initiate the contraction response to mechanical stimulation. The contractions that occur in response to light can be eliminated by removal of the hypostome and tentacles. The isolated hypostome and tentacle region remains light sensitive.

Perhaps the most complex behavior pattern of hydra is the feeding reaction. Small crustaceans such as daphnia and cyclops or small annelids are the hydra's usual prey and stimulate the feeding reaction, although hydra will ingest algae, protozoa, other hydra, planaria (see Forrest, 1962), and even mud (Wilson, 1891). The feeding response can also be stimulated in the absence of food by glutathione (Loomis, 1955; Lenhoff, 1961) and other substances including nicotinic acid, urea, glucosamine, and citric acid (Forrest, 1962). Burnett et al. (1963) have suggested that a feeding hormone responsible for the feeding response is released by nematocysts upon discharge. The hydra's prey is impaled or entwined by the nematocysts. The tentacles bearing the food contract and bend toward the mouth which is open and distended; they are dragged across the mouth, releasing some of the food, or they may extend a short distance into the digestive cavity. The open mouth glides or extends over food in its vicinity which is then forced into the digestive cavity by closure of the mouth and circular contractions of the upper part of the body. The hydra may become quite engorged with food and, after feeding, bends and twists in different directions and drags the tentacles along the substrate detaching unconsumed food. Some time after feeding, the hydra may contract violently, expelling undigestible material such as chitinous exoskeletons through the mouth.

The electrical correlates of some of these activities have been studied in hydra with the use of microelectrodes by Passano and McCullough (1964, 1965). Two pacemaker systems were described. The contraction–burst system arises from a site in the region of the hypostome and base of the tentacles. In this system a large, slow compound potential is conducted at a rate of 15 centimeters per second and precedes the periodic body contractions. The second system is the rhythmic potential system which has no obvious behavioral correlate. These potentials arise from multiple loci near the base. The activity of these systems appears to arise spontaneously in the absence of external stimuli and have their own through-conducting systems throughout and running the length of the hydra. The two systems are affected, however, by stimuli such as light and appear to interact, each altering the other's activity. McCullough (1965) suggests that the basic mechanisms of coelenterate behavior are endogenous rhythmicity and multiple conducting systems and the integrated responses of effectors controlled by the interdependent conducting systems.

Although potentials have been recorded in hydra, they do not necessarily arise in, or are conducted by, neurons. In other coelenterates, non-nervous conduction may occur (Josephson, 1965; Mackie, 1965). Josephson and Macklin (1967) detected a maintained electrical potential of 15 to 40 millivolts across the two epithelial layers of hydra, the inside of the animal being positive. Negative (depolarizing) spikes were recorded spontaneously and sometimes in response to depolarizing current pulses. The large size of these contraction pulses and the fact that the potential gradient was perpendicular to the principal axis of the neurons indicated that these potentials may be epithelial in origin. The possibility that the epithelial responses are triggered by activity conducted along nerve cells was not ruled out. In the siphonophore *Hippopodius,* however, all-or-none conduction occurs in the exumbrellar epithelium which lacks nervous elements (Mackie, 1965). Specialized intercellular junctions occur between epithelio-muscular cells of hydra (Wood, 1959; Lentz and Barrnett, 1965a), and these might allow epithelial conduction. Mackie (1965) has emphasized that epithelial and epitheliomyoid conduction may supplement nervous conduction and thus be of significance in the behavior of Hydrozoa.

The nervous system of hydra exhibits features of the nerve net, characteristic of coelenterates in general. For example, if a hydra is split longitudinally but left connected by a piece of body wall, stimulation of one side results in contraction of both (Semal-Van Gansen, 1952). Thus conduction is diffuse, indicating many available pathways. As Bullock and Horridge (1965a) have emphasized, the concept of a nerve net does not imply protoplasmic continuity of nerve fibers but rather a widespread arrangement of fibers that allows diffuse conduction of nervous excitation. Although some workers have felt that there is protoplasmic continuity throughout the nerve net, the nervous system of hydra appears to be composed of separate individual cells whose processes do not fuse (Lentz and Barrnett, 1965a). Because the nervous system of hydra is composed of individual cells and still behaves as a net, it might be expected that the anatomical discontinuities (synapses) are unpolarized. Furthermore, decremental conduction does not necessarily occur as sometimes assumed (Bullock and Horridge, 1965a). Because progressively weaker stimuli elicit progressively smaller areas of response, it was thought that decremental spread of excitation occurred in the nerve net. However, since the cells are not continuous, all-or-none nerve impulses could occur in hydra as in other coelenterates.

It is not known whether hydra are capable of learning. It has been pointed out, however, that simple organisms with a small number of cells are favorable systems for determining the morphological, histochemical, and biochemical correlates of learning and memory (Applewhite and Morowitz, 1966). From this standpoint, hydra might be an appropriate animal for investigation of learning mechanisms. On the other hand, sessile coelenterates could be poor subjects for studies of learning because they have relatively few simple reflexes, respond slowly, and show complex behavior patterns (Ross, 1965). It is of considerable interest, however, to know whether these animals that lack a central nervous system are capable of learning. There is evidence for habituation in hydra (Rushforth, 1965), and other coelenterates show adaptive changes in behavior (Robson, 1965). Ross (1965) has performed some experiments on sea anemones which indicate these animals can be conditioned. Coupling of a specific mouth-opening response to food with electrical stimulation in *Metridium,* in a few cases, resulted in apparent conditioning in which the mouth opened in response to electrical stimulation alone. Another series of experiments was performed by Ross on *Stomphia coccinea,* which releases its hold on the substrate and engages in swimming movements in response to two species of starfish, *Dermasterias imbricata* and *Hippasteria spinosa. Dermasterias* stimulus was paired with a sharp mechanical stimulus to the base of the anemone which caused it to close up and contract. After several trials, the animals did not swim to the first application of *Dermasterias* stimulus but instead contracted. Since mechanical stimulation is a punishing treatment, this type of learning is negative conditioning or conditioned inhibition. Animals that had fewer conditioning trials swam after a fewer number of tests than those with more conditioning, indicating a reinforcement–extinction relationship.

In regard to learning, it may be significant that the nervous system of hydra is not static and fixed but is replaced in accordance with the growth processes of the animal. A growth region exists below the hypostome where cells are constantly dividing and differentiating. These cells migrate distally and proximally to replace loss of aged and dying cells at the tips of the tentacles and the base. By means of this process, all the cells of hydra are renewed about every 45 days (Brien and Reniers-Decoen, 1949). Brien (1960) considers the hydra to be immortal and ponders the effect of constant cell renewal on the individuality and personality of the animal. Neurons are included in the growth process, and their differentiation from interstitial cells has been observed (Lentz, 1965a).

Nematocysts are complex organoids contained within cnidoblasts which are most abundant in the tentacles. Upon appropriate stimulation, a thread is discharged which pierces and kills the hydra's prey (see review by Picken and Skaer, 1966). It is usually assumed that nematocysts are independent effectors, responding to a combination of mechanical and chemical stimuli (Pantin, 1942; Jones, 1947). It was first believed that control of nematocyst discharge depends on the nervous system (Chun, 1881; Lendenfeld, 1887; Murbach, 1893). Mechanical stimulation or chemical stimulation alone does not produce discharge (except for irritating substances such as acid). A combination of two, however, results in discharge. Among the chemical agents that produce discharge in the presence of mechanical stimulation are the substrates (e.g. glucose-6 phosphate, adenosine triphosphate) of enzymes (glucose-6 phosphatase, adenosine triphosphatase) present in the nematocysts or cnidocils (Lentz and Barrnett, 1961, 1962). Inhibitors of the enzymes are effective in blocking discharge. In addition to these substances, several neurohumoral transmitters (acetylcholine, epinephrine, norepinephrine, 5-hydroxytryptamine, and histamine) produce discharge in the presence of mechanical stimulation, implicating the nervous system in control of nematocyst discharge. Because inhibition of the nervous system does not block the effect of chemical stimulation, while enzyme inhibitors will block nervous stimulation, both the nervous system and chemical stimuli may act through metabolic enzyme systems that represent the intrinsic final common pathway. It appears that a combination of mechanical and chemical stimulation is necessary for discharge. The effect of the nervous system could be to lower the threshold of excitability to these stimuli. The neural influence may arise from ganglion and sensory cells and could be mediated by neurohumors or the nerve fibers innervating cnidoblasts.

Nematocysts can still be considered independent effectors because, consistent with Parker's definition, the nervous system seems only to alter the response to external stimuli. Nervous activity probably does not initiate nematocyst discharge. Some other indirect evidence indicates that coelenterates have some control, possibly nervous, over discharge. Fully fed hydra, contracted hydra, or hydra distended with glass beads do not discharge as many nematocysts as unfed relaxed hydra (Hyman, 1940; Burnett et al., 1960). In *Calliactis,* the threshold to discharge is raised when the pedal disk is fastened to a whelk shell (Davenport et al., 1961). A symbiotic relationship exists between sea anemones of the families Stoichactiidae and Actiniidae and fishes of the family Pomacentridae. Of the several explanations concerning inhibition of the anemone's nematocysts (see Mariscal, 1966), one is that neural control in response to a stimulus provided by the fish may be capable of inhibiting or raising the threshold of nematocyst discharge.

The nervous system of hydra is necessary for regeneration. Dilute solutions of a variety of inhibitory neuropharmacological agents (e.g. physostigmine, DFP, atropine, reserpine) inhibit in varying degrees regeneration and attainment of normal anatomical form by transected hydra (Lentz and Barrnett, 1963). These results indicate the nervous system is necessary for the attainment of normal form of hydra, since, when it is inhibited, regeneration does not occur. Fine structural studies revealed that neurosecretory granules are released during regeneration (Lentz, 1965b). During regeneration of transected hydra, dense membrane-bounded granules disappear from the perikaryon of neurosecretory cells and accumulate in the distal processes. Granules are released by

Fig. 6. Nervous system of hydra as demonstrated by methylene blue stain. The cells are bipolar or multipolar and contain vesicles or intensely staining droplets. The nerve fibers cross and branch to form the nerve net, but there is no protoplasmic continuity between cells. The neurites may contain stained droplets along their length or have bulbous enlargements containing vesicles and droplets. The neurites may end in small knobs or larger distensions. Two terminations are shown on cnidoblasts.

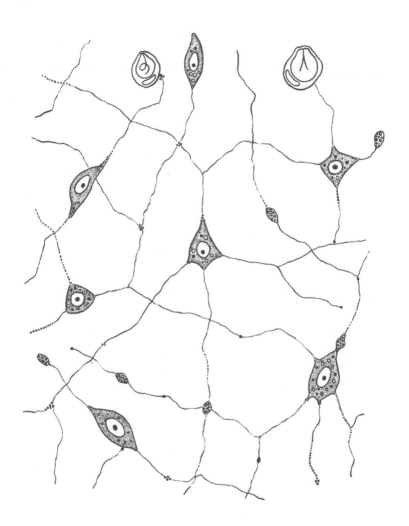

fusion of the vesicle membrane with the plasma membrane of the neurite to form a stoma through which the vesicle contents can diffuse. It appears that neurosecretory substance released at nerve terminals plays a significant role in the regulation of growth and differentiation in the regenerating hydra.

In order to test this hypothesis further, neurosecretory granules were isolated by differential centrifugation (Lentz, 1965c). Excised midsegments of hydra that were exposed to the fraction containing neurosecretory granules developed supernumerary heads as they regenerated. Normally, a head (hypostome and tentacles) develops from the distal portion of a midsegment and a base from the proximal end. Regenerates treated with the fraction containing neurosecretory granules may develop two heads distally, a head distally and another from the body, or heads distally and proximally; or they may develop into bizarre shapes. These results substantiate the hypothesis that a growth-stimulating or form-regulating factor is present in the neurosecretory granules. It appears that at least one function of this material is stimulation of head formation because, during regeneration, release of neurosecretory granules occurs at the site where a head is to appear, and exogenous administration of isolated granules induces formation of supernumerary heads. Thus, neurosecretory granules containing form-regulating substances, when transported to specific sites by neurites, may be responsible for maintenance of form in normal hydra and for the acquisition of form in regenerating hydra.

Lesh and Burnett (1966) have identified a chemical factor that controls form and polarity in hydra. This substance is probably the same as that contained within the neurosecretory granules. As with isolated neurosecretory granules, the polarizing factor induces supernumerary head formation in regenerating annuli. The active material may be a peptide, because it is dialyzable and digested by trypsin. Lesh and Burnett postulate that this factor directs the differentiation of interstitial cells which in the wound area bind the active substance and differentiate into nerve cells. These nerves in turn produce additional amounts of material that controls the development of the remaining interstitial cells. The authors suggested that this factor controls the normal form of hydra by regulating the direction of differentiation of interstitial cells.

Burnett and Diehl (1964b) found that neurosecretory substances undergo cyclic changes in response to day length. They suggest that light receptors stimulate the neurosecretory cells to release their growth-stimulating substance which promotes interstitial cell differentiation in the somatic direction. In sexual hydra, neurons are devoid of secretory droplets, and interstitial cells differentiate into gametes. Normally, hydra enter the sexual state in the fall months when daylight is decreasing. Thus, during periods of shorter daylight, the neurosecretory cells may be stimulated less. Release of growth-promoting substance then decreases, allowing interstitial cell differentiation into gametes.

Histology

The nervous system of hydra has been studied extensively by methylene blue stain techniques (Hadzi, 1909; McConnell, 1932; Spangenberg and Ham, 1960; Burnett and Diehl, 1964a). Beautiful preparations can be obtained with the vital staining technique of McConnell (1932) (reviewed by Burnett and Diehl, 1964a), which employs a methylene blue stain reduced to its leuco form by Rongalit.* Ganglion cells and sensory cells are demonstrated by this method (Fig. 6). Nerve cells are most abundant in the proximal fourth of the tentacles, hypostome, and base. Most occur

---

*Methylene blue stain for nervous tissue (after McConnell, 1932; Burnett and Diehl, 1964a).
1. Add 1 cc normal HCl to 100 cc 0.5% methylene blue.
2. Add 20 cc 15% sodium formaldehyde sulfoxylate (Rongalit) to above solution.
3. Warm *slowly* (do not boil). Stir or swirl while heating. The blue solution will change to green and then become clear with a yellow precipitate. When the solution becomes clear, remove immediately from flame and cool.
4. Let stand 24–36 hours before using. Stain is good for 1–2 weeks.
5. Add 1–2 cc stain to 50 cc water or add small amount of stain to a drop of water on a slide. Stain animals for 1–60 minutes until neural elements are demonstrated.

Fig. 7. Ganglion cell of hydra. The cell is elongated, with a neurite extending from one pole. The oval nucleus contains a nucleolus. Well-developed Golgi complexes (G) occur adjacent to the nuclear envelope. This organelle is composed of stacks of membranous lamellae and small vesicles. Mitochondria occur in the cytoplasm, often near the ends of the Golgi lamellae. A few rough-surfaced cisternae of endoplasmic reticulum are present, and free ribosomes are numerous in some cells. A few glycogen granules are present. Microtubules (MT) extend from the region of the nucleus in the longitudinal axis of the cell into the neurite.

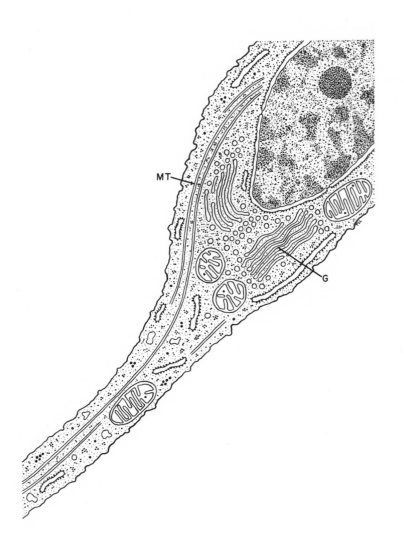

in the epidermis although a few are present in the gastrodermis. Ganglion cells are situated at the bases of the epitheliomuscular cells above the muscular processes and parallel to the surface (Figs. 5, 6). Ganglion cells are bipolar or multipolar with up to seven processes. In methylene blue preparations, the axons of ganglion cells branch and intercommunicate to give the appearance of an anatomically continuous net. The neurites contain small stained droplets which may accumulate in swellings or bulbous enlargements along the fibers. Neurites of ganglion cells terminate at the bases of cnidoblasts, on the muscular processes of epitheliomuscular cells, and on other ganglion cells. Many neurites appear to end blindly in the intercellular spaces.

Two types of ganglion cells have been described (Burnett and Diehl, 1964a); one is usually bipolar with short axons, the other is bipolar or multipolar with long axons. These types occur in approximately equal numbers. Ganglion cells of the base are the bipolar type with short axons. The axons in the base are devoid of droplets.

Sensory cells are thin elongated cells perpendicular to the surface of the animal (Fig. 5). The distal surface of the cell contains one to five sensory hairs or cilia projecting above the epidermal surface or ends as a blunt projection. Proximally, an axon extends into the nerve net or terminates on a ganglion cell. Neurosensory cells are deeply situated ganglion cells that have a cilium extending to the surface of the epidermis.

Neurochemistry

The following enzymes and neurohumors have been identified with histochemical methods in the nervous system of hydra: acetylcholinesterase (Lentz and Barrnett, 1961), monoamine oxidase (Lentz, 1966a), epinephrine, norepinephrine, and 5-hydroxytryptamine (Wood and Lentz, 1964). Of these substances, acetylcholinesterase occurs in the largest number of cells and shows the most intense staining. This enzyme is localized to small granules within ganglion cells and their processes. Stained ganglion cells appear slightly more numerous in the hypostome. Sensory cells are also reactive, although not so intensely as ganglion cells. The reactive neurites of ganglion cells terminate on reactive grape-like clusters of granules on the surfaces of cnidoblasts and epithelio-muscular cells. Monoamine oxidase activity occurs in only a few ganglion cells and neurites, and most of these are present in the peduncle and base.

Catecholamines and 5-hydroxytryptamine occur in ganglion cells, sensory cells, and their neurites. The stained cells are most numerous in the hypostome but not as many can be demonstrated as with acetylcholinesterase. Staining occurs in small granules and vesicles.

Neurosecretory cells have been identified by staining with Gabe's paraldehyde fuchsin method, periodic acid Schiff after salivary digestion, methylene blue, and toluidine blue (Burnett et al., 1964). These techniques reveal small droplets in the cell bodies and within the neurites, especially in the hypostome and proximal regions of the tentacles. Nerve cells with extremely large axons containing stained droplets occur at the bases of the tentacles.

Neurohumors have been identified in coelenterates by biochem-ical methods. Reports on the occurrence of acetylcholine are con-flicting. Bacq (1935, 1947) was unable to identify acetylcholine in coelenterates. Mitropolitanskaya (1941) found none in hydra but detected it in *Actinia equina* and *Aurelia aurita*. Bullock and Nachmansohn (1942) found acetylcholinesterase in *Tubularia, Metridium,* and *Sagartia* and small amounts in *Cyanea* and *Aure-lia*. Mitropolitanskaya identified this enzyme in *Hydra* and *Actinia* and Augustinsson (1948) identified it in *Sagartia*. Östlund (1954) stated that sea anemones contain a catecholamine but that it is not epinephrine, norepinephrine, or 5-hydroxytryptamine. Welsh and Moorhead (1960) found relatively large amounts of 5-hy-droxytryptamine in *Hydra, Sagartia,* and *Metridium*. In the last species, 5-hydroxytryptamine occurred in highest concentration in the acontia. Mathias et al. (1960) also found 5-hydroxy-

Fig. 8. Neurosecretory cell of hydra. The cell contains large, dense, membrane-bounded granules (NSG) of moderate density. The granules occur in the perikaryon and processes of the spindle-shaped cell. Some granules are in close relation to the Golgi apparatus (G), and one is shown within the end of a dilated lamella. The Golgi apparatus is well developed, consisting of membranous lamellae and small vesicles (V). Some of the Golgi elements contain dense material. A mass of moderately dense substance occurs in the cytoplasm adjacent to the Golgi apparatus. Other cytoplasmic structures are mitochondria, endoplasmic reticulum, ribosomes, and glycogen granules (Gly). Microtubules (MT) extend into the neurite. The nucleus (N) contains a nucleolus.

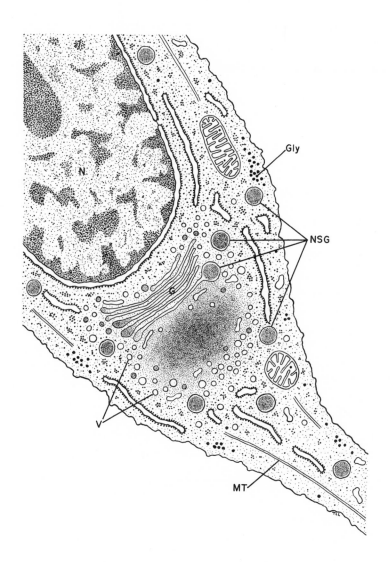

tryptamine in *Calliactis* as well as other pharmacologically active substances in sea anemones and *Physalia*.

Dahl et al. (1963) localized monoamines in sea anemones under fluorescence microscopy in tissues treated with formaldehyde gas. In the tentacular apparatus, they observed neurons in the ectodermal epithelium. A fiber containing varicosities and a subterminal expansion extends from the cell body to the surface, and another fiber extends basally to the ectodermal muscular layer.

Fine structure

Nerve cells in hydra have been studied with the electron microscope (Lentz and Barrnett, 1965a; Lentz 1965a, 1966a). As expected from the light microscopic findings, nerve cells are located at the bases of epitheliomuscular cells above the muscular processes and are most abundant in the hypostome. Most neurons occur in the epidermis; very few were observed in the gastrodermis. Several cell types and their processes comprise the nervous system. Besides ganglion, sensory, and neurosensory cells described at the light microscopic level, some cells were found to contain dense granules and are considered to be neurosecretory cells.

Ganglion cells are small bipolar cells at the bases of epitheliomuscular cells (Fig. 7). No sheath or satellite cells are present, although the ganglion cells are usually closely enveloped by thin processes of epitheliomuscular cells. The plasma membrane is thrown up into many small crests and indentations. The centrally situated nucleus is bounded by an envelope containing pores. The chromatin material is evenly distributed.

A nucleolus may be present, containing dense granules 150 A in diameter. A few rough-surfaced cisternae of endoplasmic reticulum occur in the cytoplasm. Free ribosomes are abundant in some cells. A few cells are nearly devoid of endoplasmic reticulum and ribosomes. Mitochondria are oval, with a few cristae extending across the organelle. Some cells contain membrane-bounded masses of dense material within which may be embedded granules or vesicles. These bodies are probably a type of lysosome. Glycogen granules and a few irregular membranous sacs are present in the perikaryon. Microtubules are numerous in ganglion cells. These structures are about 220 A in diameter and extend parallel to the long axis of the cell. They appear to originate in the vicinity of the nucleus, sometimes in close proximity to nuclear pores, and sweep around the nucleus, extending out into the processes.

Ganglion cells possess one or more complex Golgi apparatus. This organelle is usually situated in the long axis of the cell and often occurs between the nucleus and a process. The Golgi is composed of a stack of membranous lamellae and small vesicles, 500 A in diameter. The vesicles are especially numerous near the ends of the Golgi lamellae and contain material of low or medium density. Occasionally, larger vesicles with moderately dense contents occur in the vicinity of the Golgi apparatus or in the perikaryon.

Neurosecretory cells are similar in structure to ganglion cells (Fig. 8). The major difference between these cells occurs in the Golgi region. The large Golgi apparatus of neurosecretory cells is near the nucleus and contains membranous lamellae and small vesicles containing moderately dense material. In addition, dense membrane-bounded granules, 1,000 to 1,200 A in diameter, occur in relation to the Golgi apparatus. On some occasions they are situated in the dilated ends of the Golgi lamellae. A mass of moderately dense amorphous material occurs adjacent to or surrounding the membranous elements of the Golgi apparatus. This material may represent unaggregated precursor of neurosecretory substance. The dense granules are scattered throughout the perikaryon of ganglion cells. Only slight variations in the size and density of the granules were noted.

It is not definitely known what the content of the dense granules is. They appear to contain a growth-stimulating substance, because hydra grow supernumerary heads when exposed to fractions containing neurosecretory granules. Moreover, it was shown that

Fig. 9. Sensory cell of hydra. This elongated cell contains an apical specialization and a basal process which extends into the nerve net. The distal end consists of an indentation of the surface membrane and a cilium or sensory hair (SH) arising from the bottom of the indentation. The ciliary fibers merge with dense material comprising the basal body (BB). Rootlets (Rt) extend downward toward the nucleus. A well-developed Golgi apparatus (G) is located above the nucleus. Small vesicles (V) occur in association with the Golgi apparatus. Mitochondria, endoplasmic reticulum, and glycogen granules (Gly) are present. Microtubules (MT) course longitudinally in the cell and its process.

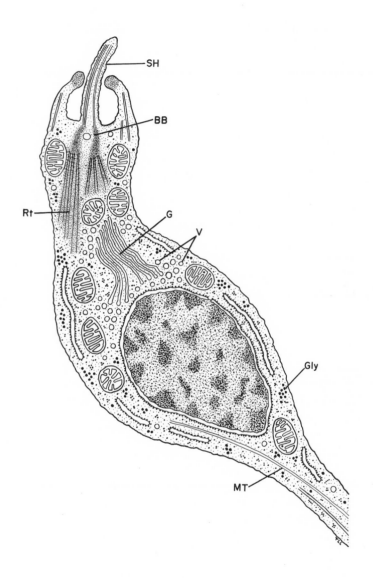

epinephrine, norepinephrine, 5-hydroxytryptamine, and acetyl-cholinesterase are present in the hydra nervous system. Catechol-amines and 5-hydroxytryptamine occur in dense granules in more evolved nervous systems. In hydra, it is not known whether these substances might occur in different granules or cells from those containing neurosecretory substance with a hormonal action, because only one type of granule was observed. If the presence of acetylcholinesterase is indicative of cholinergic neurons, these would most likely be the ganglion cells, because they contain small vesicles resembling the synaptic vesicles of vertebrates that contain acetylcholine (DeRobertis et al., 1963; Whittaker et al., 1964). Small vesicles also occur in neurosecretory cells.

Microtubules (neurotubules) are prominent in hydra neurons, as in most other nerve cells. Several functions have been suggested for microtubules, but there appears to be a consistent relationship between microtubules and both maintenance of form (Tilney and Porter, 1965; Gibbins et al., 1966) and cytoplasmic streaming (Ledbetter and Porter, 1963; Tilney and Porter, 1965; Lentz, 1967a). In nerve cells, microtubules could be associated with the flow of axoplasm or vesicles down the axon and also provide support to the elongated cell process (Lentz, 1967c). During regeneration of transected hydra, microtubules increase in number, perhaps accounting for the accumulation of neurosecretory granules in the nerve endings prior to their release (Lentz, 1965b, 1966a).

Ganglion and neurosecretory cells are very similar, differing only in the size and density of contents of the vesicles originating in the Golgi region. It is possible that they represent the same basic cell type, which, under certain physiological demands, is capable of elaborating neurosecretory granules. The two cells might also have separate functions but represent the same cell line in different functional states or stages of differentiation.

Sensory cells are situated between the apices of epithelio-muscular or digestive cells and usually contain an apical special-ization (Fig. 9). These cells do not differ from ganglion cells in cytoplasmic content. Microtubules extend in the long axis of the cell. The Golgi apparatus is usually located in a supranuclear position and contains small vesicles, 500 A in diameter.

The characteristic feature of sensory cells is their apical specialization which is usually a cilium. A single cilium arises from a surface indentation produced by infolding of the apical plasma membrane. The cilium contains nine pairs of peripheral rods, but up to five central rods have been observed. The rods merge with dense material comprising the basal body at the base of the cilium. From the basal body, filaments or rootlets splay out into the apical cytoplasm. A few sensory cells contain a bulb-shaped apical process lacking a cilium.

As described previously, hydra are sensitive to a number of stimuli including light. Since one of the basic types of photoreceptors found in higher organisms is of ciliary origin, the sensory cell with the specialized cilium could be a photoreceptor. A photoreceptor cell in the hydromedusan *Polyorchis penicillatus* contains a cilium but is otherwise much more elaborately specialized than the sensory cell of hydra (Eakin and Westfall, 1962). It is not known whether this cell is sensitive to more than one type of stimulus, but Rushforth et al. (1963) presented evidence that hydra have separate receptors for light and mechanical stimuli.

Neurosensory cells are similar to ganglion or neurosecretory cells but in addition possess a cilium (Fig. 10). Clear vesicles or dense granules are present in the cytoplasm. The cilium arises from the perikaryon or from a neurite and extends toward the surface. As in superficially located sensory cells, the cilium protrudes from an indentation of the plasma membrane of the cell. The space produced by the indentation is usually dilated around the base of the cilium. A basal body is present and sometimes long striated rootlets extend toward the nucleus. In some cells, numbers of finger-like processes project downward into the dilated space at the base of the cilium. These processes or evaginations contain a central core or rod.

The processes or neurites of the nerve cells arise from the perikaryon and extend above the muscular processes (Fig. 7). The

61

Fig. 10. Neurosensory cell of hydra. The perikaryon of neurosensory cells is the same as that of ganglion or neurosecretory cells. In the cell illustrated, a Golgi apparatus (G) with associated dense granules occurs adjacent to the nucleus (N). Mitochondria, microtubules, a few cisternae of rough-surfaced endoplasmic reticulum, ribosomes, and glycogen granules occur in the cytoplasm. The cell contains a blunt protrusion from which a cilium (Ci) extends toward the surface. The cilium arises from the bottom of an indentation on the surface of the cytoplasmic process. Striated rootlets (Rt) extend from the basal body (BB) to the perinuclear region.

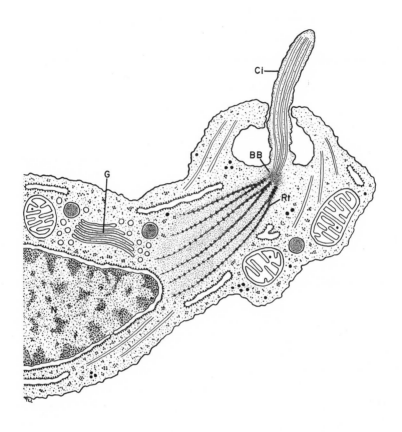

fibers are unmyelinated but are often surrounded by processes of epitheliomuscular cells. Ribosomes, glycogen granules, and mitochondria occur in the processes. Microtubules extend their entire length. Dense granules, vesicles, and membranous sacs with lucent contents are present. Many neurites contain bulbous enlargements along their length. These swellings are often filled with vesicles or granules. Giant fibers filled with neurosecretory granules occur at the bases of the tentacles and probably correspond to the large fibers observed in methylene blue preparations (Burnett et al., 1964).

Terminations of neurites occur in intercellular spaces, on cnidoblasts, on the muscular processes of epitheliomuscular cells, and on other neurons (Fig. 11).* Microtubules extend into the nerve endings and sometimes come into close approximation to the distal plasma membrane. Mitochondria, ribosomes, glycogen granules, and small vesicles occur in the terminations. There are dense neurosecretory granules in heavy concentrations. Some of the vesicles and granules are adjacent to the plasma membrane of the neurite. Large, dilated, empty sacs occur in the same areas.

Some neurites appear to terminate in large intercellular spaces. When neurites end near other cells, the intervening space is highly irregular, varying from 200 to 500 A across. Neither the membrane of the neurite nor the target cell is thickened in the region of contact. Although vesicles and granules are abundant in the nerve endings and a few occur adjacent to the plasma membrane, they do not show a preferentially heavy concentration in the region of contact. These nerve endings, therefore, are relatively unspecialized and do not contain all the features of synapses.

Since the hydra nervous system is composed of separate cells

---

*I believe that functional communication occurs between these nerve endings and other neurons or effectors. In this case, the nerve endings near other cell types may be considered synapses in a general sense. There are few morphological specializations, however, that might indicate the precise region of transmission. Because there do not appear to be sufficient anatomical criteria to exclude nonsynaptic interactions beween cells, the nerve endings in hydra have not been referred to as synapses (Lentz, 1966a).

instead of a fused net and contains neurotransmitters, transmission is probably chemical rather than electrical. The unpolarized nature of the nerve endings and the sometimes large spaces separating the nerve ending from the effector indicate that this system might be slow acting, variable, and diffuse. Although this is compatible with some of the physiological and behavioral observations such as local responses, it does not explain the rapid through-conducting motor pathways. Through-conduction with little or no synaptic delay could occur in a synaptic nerve net if the synapses consisted of regions of close membrane apposition or fusion. This type of synapse does not occur in hydra, although it could be present in other coelenterates. In the absence of these specialized synapses in hydra, through-conduction could occur in individual neurons with many branching and very long processes. Although it is not known if individual nerve fibers extend the length of the animal, it appears from methylene blue preparations that some may extend at least a quarter to a third of this distance. The alternative explanation is that through-conduction does not occur in nerves but in other cells such as epitheliomuscular cells. The findings of Josephson and Macklin (1967) indicate that spike-like electrical potentials are not only conducted by epithelial cells but may originate in them as well.

Jha and Mackie (1967) have performed fine-structural studies on the nervous systems of several other hydrozoans, including the hydroid *Cordylophora,* the medusae of *Sarsia* and *Euphysa,* and the nectophores of the siphonophore *Nanomia.* An important aspect of their work is the careful correlation of light and electron microscopic recognition of neurons. Nerve cells in these hydrozoans seem to be basically similar in structure to those of hydra. Special mention, however, should be made of sensory hairs and vesicular structures.

Sensory hairs were observed on many neurons and have the usual $9 + 2$ pattern of tubules and a single striated rootlet in the cytoplasm. This type of sensory hair was thought most likely to be mechanoreceptive. In the ocellus of *Sarsia,* the membrane of the

65

Fig. 11. Types of nerve endings in *Hydra littoralis*. The neurites occur at the bases of epitheliomuscular cells (EmC) above the muscular processes (MP) which contain thick and thin myofilaments and overlie the mesoglea (Me). The nerve endings contain membrane-bounded neurosecretory granules (NSG) and small vesicles (V). Larger dilated vesicles or sacs are devoid of dense contents. Microtubules (MT) are situated in the long axis of the processes. Mitochondria, endoplasmic reticulum, ribosomes, and glycogen granules occur in the endings. Four nerve terminations are illustrated. Neurite 1 ends in an intercellular space. Neurite 2 terminates adjacent to a cnidoblast cell containing a nematocyst (Nt). Neurite 3 is dilated terminally and occurs next to a ganglion cell (GC). Neurite 4 ends as a bulbous enlargement on a muscular process. In all cases a space of irregular diameter separates the nerve from other cell types, and there is no localized thickening of the plasma membrane. The granules and vesicles may occur adjacent to the plasma membrane of the neurite but are not concentrated in localized regions.

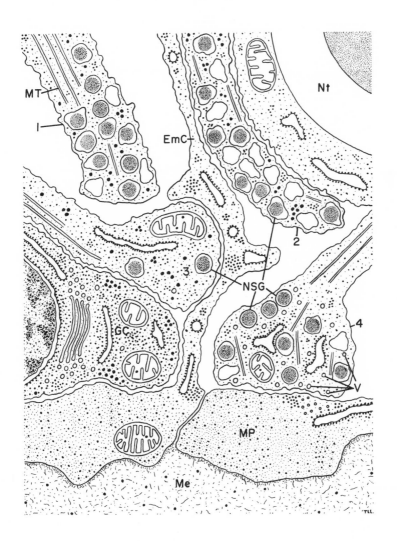

sensory cell cilium is drawn out into numerous villous processes. This cell is photosensitive and similar to the photoreceptor of *Polyorchis* described by Eakin and Westfall (1962).

Two basic types of vesicular inclusions were described by Jha and Mackie. Some vesicles are small with clear contents while others are larger and contain dense material. The latter seem to be the same as the neurosecretory granules of hydra. These structures occur in the nerve cell bodies and their processes. Nothing comparable to a synapse was identified in *Cordylophora, Euphysa,* or *Nanomia,* but in *Sarsia* two types were observed. In some adjacent nerve cell processes, vesicles occur on both sides of the junction. These synapses are similar to those described in *Cyanea* by Horridge and Mackay (1962). The second type, however, is structurally polarized with vesicles occurring on only one side of the junction. The latter is the first description of a polarized synapse in coelenterates.

In extending the number of coelenterate species that have been examined, the studies of Jha and Mackie (1967) reveal a basic similarity in coelenterate neurons. Nerve cells in this phylum have some of the structural characteristics of neurons in higher animals. Interneuronal relationships such as synapses, on the other hand, are not nearly so specialized. In general, the nervous system of coelenterates is less complicated than the nervous systems of higher animals, and of existing and definitely identifiable systems, therefore, it may most closely resemble a primitive or an elementary nervous system.

4
THE NERVOUS SYSTEM
OF PLANARIA

Introduction

The Platyhelminthes include the free-living Turbellaria or flat-worms and the parasitic Trematoda and Cestoda. The Turbellaria, which are considered here, are flattened worms occurring in fresh water, salt water, or on land. They are bilaterally symmetrical and the anterior end is differentiated into a head. Flatworms have well-defined organ systems and are covered by a single-layered epidermis. Cilia occur in the epidermis, especially ventrally, and serve as a means of locomotion. Rhabdoids, which include the rhabdites, are organoids that produce a slimy coating around the animal when discharged. Most forms of Turbellaria have a mouth and pharynx leading to an intestine lined by a gastrodermis composed of phagocytic and glandular cells. A solid mesenchyme occupies the space between epidermis and gastrodermis and contains the organ systems. Fixed cells with long processes, neoblasts or formative cells (wandering undifferentiated cells), and several types of gland and pigment cells occur in the mesenchyme. Muscle fibers are present in the mesenchyme and also in a subepidermal layer. The excretory system consists of flame bulbs and proto-nephridial tubules. The reproductive system is complex and includes gonads, yolk or vitelline glands, and a copulatory apparatus. In all except the most primitive Turbellaria, the nervous system has sunk into the mesenchyme.

Many of the nerve cells comprising the nervous system of the flatworms closely resemble those of vertebrate nervous systems. From a structural standpoint, this system perhaps should not be considered primitive. However, because many features of the evolved system make their first appearance in this phylum, it deserves consideration and comparison with the less complex forms. The nervous system of some of the primitive Turbellaria (Acoela, Alleocoela) bears some resemblance to that of coelenterates and consists of a network of cells and fibers at the base of the epidermis above the muscular processes of epithelial cells. There is a tendency for concentration of neurons anteriorly and the formation of longitudinal strands in these lower forms. In more advanced Turbellaria, the nervous system, which has sunk into the mesenchyme, forms a submuscular plexus. In addition, neurons are concentrated into several longitudinal nerve cords and anterior cerebral ganglia or brain. Turbellaria are well supplied with sensory cells and organs, including tactile cells or tangoreceptors, ciliated pits and grooves containing chemoreceptors, rheoreceptive cells, frontal organ, statocyst, and ocelli. The familiar triclad flatworms exhibit most of the characteristics of the phylum and will be used as an example of this level of organization.

Physiology

Movement of small Turbellaria occurs by anteroposterior ciliary waves independent of the nervous system. Larger forms move by means of muscular waves passing from front to back. Other types of locomotion include swimming by means of body undulations and leech-like movement by alternate attachment of adhesive organs. Turbellaria are capable of twisting and bending portions of the body, righting, back and forth swinging of the head, and turning movements.

Flatworms are carnivorous and most will eat any kind of animal prey which they grasp with the anterior end and enfold. Adhesive organs or areas hold the prey and enable the worm to adhere to the substrate. The pharynx is protruded and inserted into the food which is either swallowed whole or drawn up in portions into the pharynx by peristaltic action.

Turbellaria show a number of tactic reactions, including chemotaxis, thigmotaxis, and phototaxis; they are readily attracted by food. When a worm is in the vicinity of diffusing juices, it usually stops, raises and swings its head about, and proceeds toward the food in a straight line. The lateral sensory organs of the head (ciliated pits, ciliated grooves, or auricles) are known to contain the chemoreceptors since removal of these areas abolishes or inhibits detection of food (Koehler, 1932; Müller, 1936). The anterior margin and the tip of the pharynx test the suitability of the food. The dorsal surface of Turbellaria is negatively thigmo-tactic. When turned on its back, a planarian exhibits a righting reaction. Most planaria avoid light, seeking the darkest areas of the environment. Undirected light results in increased activity of planaria, which does not depend on the cephalic eyes. Negative response to directed light depends on the eyes to some extent because it affects only the retinal cells whose long axes are parallel to the light rays (Taliaferro, 1920). Planarians living in running water exhibit positive rheotaxis (moving upstream). They exhibit negative geotaxis (rising to the surface) when oxygen is low and postive geotaxis when oxygen is adequate. The statocyst may be responsible for the geotactic reactions. Planarians also are sensitive to temperature changes (thermotaxis) and electric current (gal-vanotaxis).

The brain seems to be necessary for some of the activities of the animal. Regeneration, learning, and locomotion may depend, to some extent, on the brain. The brain may also maintain a level of excitability and spontaneity (Robertson, 1928; Bullock and Horridge, 1965a).

There is some evidence that planaria are capable of learning (see reviews by Jacobson, 1963, 1965, and McConnell, 1965). Hovey (1929) exposed the polyclad *Leptoplana* to light for 5 minutes and then to darkness for 30 minutes. When the animal thrust its head forward at the onset of illumination, the tip of its head was touched with a rod, causing the snout to retract. Each time the worm attempted to advance, it was touched. As the trials progressed, fewer touches were required to induce the animal to

Fig. 12. Nervous system of a triclad turbellarian. The pair of cephalic ganglia joined by a broad commissure and comprising the brain are located in the anterior or head portion of the animal. Short optic nerves extend dorsally from the brain to the pigmented cup of the eye. Nerves radiate from the brain to all regions of the head, the auricles being especially well supplied. A pair of longitudinal nerve cords extends posteriorly from the brain. At intervals the cords are united by transverse commissures. Lateral nerves extend from the cords. The lateral nerve bundles branch near the edge of the animal and are continuous with a fine marginal plexus representing the submuscular nerve plexus.

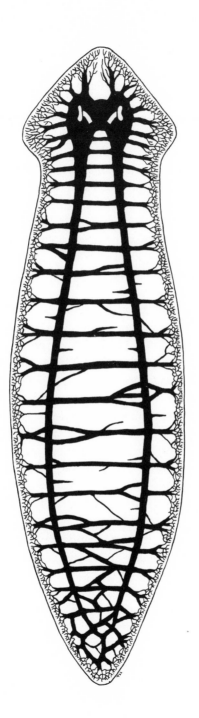

remain motionless during light stimulation. The planaria did not acquire the conditioned responses after the brain was removed. Thompson and McConnell (1955) conditioned *Dugesia* to paired light and electric shock. The shock produced sharp turning of the head or longitudinal contraction of the body. The frequency of turning and contraction at the onset of light increased significantly with training. Studies of the effects of regeneration upon retention of the conditioned response revealed that regenerated head and tail sections of transected planaria retained the response to a degree equal to retention by uncut animals (McConnell et al., 1959). Retention by the tail section is surprising, and it was suggested that the brain is necessary for the acquisition but not retention of a conditioned response.

Experiments have been performed to determine the cellular mechanisms of information storage. McConnell et al. (1961) reported that planaria which had ingested portions of trained worms showed a higher incidence of conditioned response than animals ingesting untrained worms. Corning and John (1961) found that the tail portions of conditioned planaria which had regenerated in ribonuclease did not retain the conditioned response. Finally, untrained planaria injected with RNA extracted from conditioned worms showed a higher level of response than worms given RNA from pseudo-conditioned (unpaired light and shock stimuli) animals (Zelman et al., 1963). These experiments have been interpreted to mean that learning and memory could depend on the modification of neuronal RNA, which, by providing the template for the elaboration of neurotransmitters, determines the postsynaptic firing or inhibition patterns. Although there has been considerable controversy over these experiments, they represent significant avenues of investigation into the chemical correlates of learning and memory and indicate that these processes can be studied to advantage in relatively simple organisms.

The nervous system of planaria is necessary for the maintenance of existing parts or for regeneration of excised portions (Child, 1904a,b, 1910; Olmsted, 1922). The cephalic ganglia seem to be most important in exerting this trophic effect. Lender (1955) has found that the brain plays a role in ocular regeneration in *Polycelis nigra*. Extracts of planarian heads are capable of inducing ocular regeneration. During posterior regeneration of *Polycelis,* neurosecretory cells become rounded, increase in number, and show greater amounts of secretory material (Lender and Klein, 1961). Tracts containing stained material are also present during posterior regeneration. Neurosecretory cells may play a role similar to those in hydra by releasing a substance that controls or regulates growth and differentiation.

Histology

The turbellarian nervous system varies considerably in different orders but usually consists of a brain, longitudinal cords and commissures, and plexuses (Fig. 12). The cerebral ganglia or brain, absent only in some primitive forms, is a bilobed mass situated anteriorly and consisting of a central bundle of fibers (neuropil) enclosed by several types of nerve cells. Nerve bundles radiate from the brain to the sensory organs of the head. Other nerves connect with the longitudinal cords. Dorsal, ventral, and lateral longitudinal nerve cords may be present. The cords are connected by transverse, circular, or dorsiventral commissures, and lateral nerves extend from them.

The larger nerves give rise to the peripheral plexuses of nerve fibers and cells (Fig. 13). The submuscular nerve plexus is extensive and is situated in the mesenchyme below the subepidermal musculature. Sensory cells and ganglion cells with branching and irregular processes occur in this layer. The processes may have bulbous enlargements along their length, and some terminate synaptically on muscle. The subepidermal plexus is much finer and delicate than the submuscular and is at the bases of the epithelial cells above the muscular layer. The cells are sparse and consist of spindle-shaped ganglion and sensory cells.

Nerve cells are unipolar, bipolar, or multipolar. Many types have been described (see Bullock and Horridge, 1965a). The

Fig. 13. Peripheral nerve plexuses of Turbellaria. A fine subepidermal plexus (SEP) is situated below the ciliated epithelial cells (EC) but above the muscle cells (MC). This layer is composed of small spindle-shaped cells with delicate processes. Some cells have distal processes extending to the surface. A superficial sensory cell in the epidermis has a distal process and a proximal fiber extending into the submuscular plexus. The submuscular layer (SMP) of nerves, in the mesenchyme (Mes) below the muscle cells, is composed of a larger number of cells, mostly bipolar or multipolar. The processes may be long, with bulbous enlargements along their length, or form terminal arborizations. Some of these cells have distal branching processes projecting into the epidermis. Rb, rhabdites; RGC, rhabdite-forming gland cell; MsC, mesenchymal cell; Nb, neoblast.

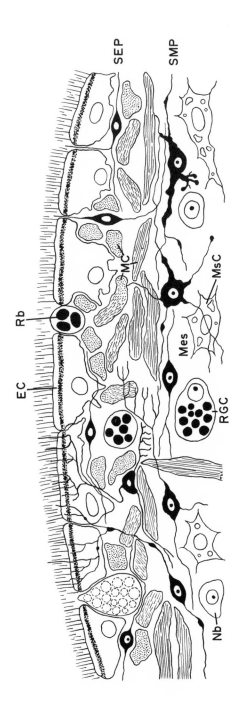

SEP

SMP

MC

Rb

EC

MsC

Mes

RGC

Nb

major differences between these cells and those of coelenterates appear to be extensive branching of the processes and presence of many unipolar neurons in planaria. Motor and internuncial neurons occur. It is generally felt that the nervous system is synaptic, although protoplasmic continuity of some fibers has been described (Gelei, 1909, in Bullock and Horridge, 1965a). Nerve cells occur in the brain and scattered along the entire plexus. Neurosecretory cells are present (Lender and Klein, 1961) and glial cells between neurons and along nerve cords have been described (Clayton, 1932; Morita and Best, 1966).

Turbellarians are abundantly supplied with sensory receptors. Single sense cells are widely scattered throughout the animal (Fig. 13). The cell bodies lie in the subepidermal and submuscular plexuses and in the epidermis. Some cells have a single distal process while others have a large number of branching distal processes (Graff, 1912–17). The processes terminate in sensory hairs or bristles extending through the epidermis and above the level of epidermal cilia. Proximally, the axon enters the subepidermal plexus, submuscular plexus, or the brain. The single primary sensory cells include tactile receptors, rheoreceptors, and chemoreceptors. Sensory organs include ciliated pits and grooves, the frontal organ, statocyst, and eye. Ciliated pits and grooves, auricular sense organs, and sensory margins are depressed regions of the epidermis, usually in the head, devoid of rhabdoids and sometimes without gland cell openings. The distal processes of bipolar neurons branch and terminate in sensory hairs at the bases of these indentations (Gelei, 1930; Müller, 1936). These structures are chemoreceptive (Müller, 1936). The frontal organ is composed of chemoreceptive sensory fibers accompanying the long necks of gland cells.

A statocyst occurs in the primitive Turbellaria and is similar to the hydrozoan statocyst. The statocyst is in or near the brain and is innervated from the brain (Westblad, 1937). It consists of a nucleated vesicle containing one or two special cells called lithocytes which enclose a round concretion or statolith. Light reception takes place in the eyes or ocelli which may occur as an anteriorly situated pair or scattered abundantly over the head region. The general body surface is sensitive to light, although the sensory receptors have not been identified. The eye is an inverse pigment-cup ocellus. Pigment cells form a cup and the distal processes of photosensitive or retinal cells enter the cup through its opening. The distal process of the bipolar retinal cell is bulb- or club-shaped with a distal striated tip or rod border. The proximal process enters the brain. In land planarians, an epithelial thickening over the mouth of the cup forms a cornea.

Neurochemistry

A striking demonstration of portions of the nervous system of *Procotyla fluviatilis* can be obtained with the thiolacetic acid–lead nitrate method for acetylcholinesterase (Lentz, unpublished observations). The subepidermal and submuscular nerve plexuses stain especially well (Fig. 14). The density of reaction product and number of cells stained indicate that most of the neurons comprising these plexuses are demonstrated. Staining in the brain is variable; the large nerve cords and epidermal sensory cells are unreactive.

The superficial subepidermal plexus is revealed with this method as a network of small bipolar or multipolar cells with fine processes (Fig. 14). The existence of this nerve layer is therefore confirmed by this method. The nerve fibers generally run in the longitudinal axis of the animal, but there are many intercommunicating branches. The cells and their processes give the appearance of a nerve net as in hydra. Most of the nerve fibers join the network, few fibers terminating outside of it. Reaction product occurs diffusely in the cells and their processes. A few small intensely stained granules occur in the cell bodies or along the fibers.

The submuscular plexus is deep to the subepidermal plexus and is intensely stained (Fig. 14). The cells are more numerous in this layer and most are bipolar. Unipolar and multipolar cells also occur. The processes extend roughly longitudinally

Fig. 14. Peripheral nerve plexuses of the turbellarian *Procotyla fluviatilis* as revealed by the localization of acetylcholinesterase activity. The subepidermal plexus (top) is a fine network composed of spindle-shaped cells and their processes. The processes tend to be oriented longitudinally, but many transverse fibers join the longitudinal fibers. This plexus gives the impression of being a continuous network but is composed of individual cells. The submuscular plexus (bottom) is much coarser than the subepidermal. The cells are larger and more numerous and their processes are of greater diameter. The cells asume a variety of shapes and some have many processes. The cells and fibers run mainly longitudinally, but there are many interconnections. The fibers are of irregular diameter with many vesicular swellings. Some of the fibers cross to give the plexus a netlike appearance and others end in swellings or terminal arborizations.

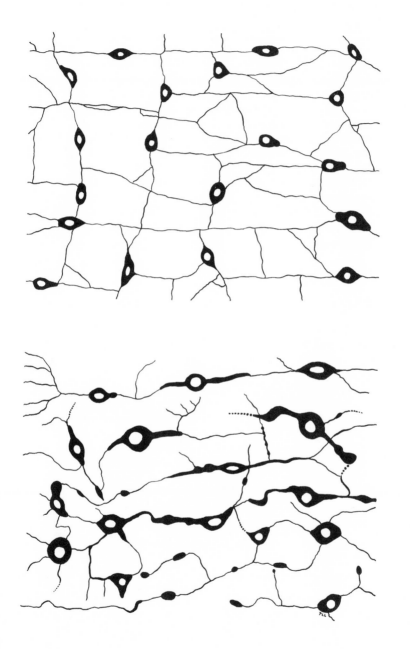

Fig. 15. Ganglion cell in the brain of the white planarian *(Procotyla).* The cell is highly irregular in shape with numerous surface projections and processes, some of which may be neurites. The cell contains a central nucleus with a prominent nucleolus. Several Golgi complexes (G) occur around the periphery of the nucleus. The Golgi apparatus is composed of stacks of membranous lamellae and many small clear vesicles (V), about 500 A in diameter. One Golgi (lower left) contains dense material within the lamellae and large vacuoles. Similar large dense bodies may be lysosomes (L). Another prominent feature of the cytoplasm are microtubules (MT) shown in both longitudinal and transverse section. These structures are very numerous, occupying much of the space between organelles. Mitochondria, cisternae of rough-surfaced endoplasmic reticulum and clusters of ribosomes (R) are present.

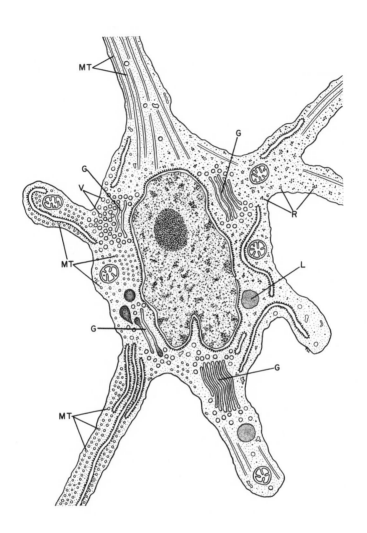

but many commissures are present. The processes have an irregular diameter with local enlargements or swellings along their length. Many of the processes appear to terminate in bulb-like swellings or fine terminal branches. A few fibers extend into the epidermis. Enzyme activity is localized to small cytoplasmic granules filling the cell bodies and distributed along the processes.

The brain is visible in some preparations. It appears as a bilobed mass of stained cells below and posterior to the eyes. Reactive nerve fibers or bundles were not seen emanating from the brain. Reaction product is localized to small cytoplasmic granules that are most abundant around the nucleus.

Lender and Klein (1961) identified neurosecretory cells in *Polycelis nigra* with a paraldehyde fuchsin stain. The stained cells are monopolar and most numerous on the posteroventral surface of the brain. These cells contain a small nucleus and an indistinct nucleolus. Stained tracts extend from the cell bodies.

Bullock and Nachmansohn (1942) found a very high concentration of acetylcholinesterase in planaria, including *Procotyla*. The histochemical experiments confirm this finding and show that the enzyme is localized to the nervous system. Welsh (1946) identified acetylcholine in planaria and found the level was greatest in the anterior third of the animal. 5-Hydroxytryptamine also occurs in planaria, but the amount in the tail is equal to or greater than that in the head portion, possibly because it may be associated with the excretory system (Welsh and Moorhead, 1960). Adrenergic neurons have been identified in turbellarians by fluorescence techniques (Dahl et al., 1963).

Fine structure

Several types of nerve cells can be identified in the nervous system of planaria with the electron microscope (Morita and Best, 1965, 1966; Oosaki and Ishii, 1965; Lentz, 1967b). Ganglion cells comprise the major portion of the brain of *Procotyla* (Fig. 15), and are tightly packed around the central neuropil. They are highly irregular in shape, with three to eight processes extending from the perikaryon. Individual cells do not appear to be surrounded by glial elements, and synapses were not observed on the soma of ganglion cells. Adjacent cells are usually separated by a space of 200 A.

A central oval or irregular nucleus contains a large nucleolus which is composed of fibrous material and a large number of dense particles, 150 A in diameter. The nuclear envelope contains a few pores. Within the cytoplasm are small mitochondria, free ribosomes, and elongated rough-surfaced cisternae of endoplasmic reticulum. The latter may occur singly or in stacks of several cisternae. Round membrane-bounded structures, 0.1 to 0.2 μ in diameter, contain granular, moderately dense material and may be a type of lysosome. Microtubules are very abundant in ganglion cells. In some cells they fill most of the available space between other organelles. Microtubules extend in the longitudinal axis of the cell from the vicinity of the nucleus out into the processes.

One or more prominent Golgi apparatus occurs in ganglion cells, adjacent to the nucleus in the long axis of the cell. A process often arises opposite the Golgi apparatus. This organelle is composed of a stack of several flattened membranous lamellae and small vesicles. The vesicles, about 475 A in diameter, are most numerous at the ends of the lamellae. The vesicles and lamellae contain material of low to medium density. Some larger vacuoles containing dense material are associated with some Golgi complexes. These granules closely resemble lysosomes.

Several processes (up to eight have been observed) may arise from ganglion cells. In some cases the region where the process arises is devoid of endoplasmic reticulum and ribosomes and resembles an axon hillock. Many processes contain mitochondria, endoplasmic reticulum, ribosomes, lysosomes, microtubules, and vesicles while others contain relatively few organelles. Processes are not enveloped by a myelin sheath or glial cells.

Oosaki and Ishii (1965) described a nerve cell in the brain of *Dugesia* that contains a large number of small vesicles (Fig. 16), 500 A in diameter; most have a content of low density. Occasion-

Fig. 16. Nerve cell occurring in the brain of *Dugesia*. Large numbers of small vesicles occur in the cytoplasm of this cell. Most of the vesicles have contents of low density, although a few are filled with moderately dense material or have a dense core. An elaborate Golgi complex (G) is adjacent to the nucleus. Cisternae of rough-surfaced endoplasmic reticulum (ER) are present and many occupy large regions of the cytoplasm from which other organelles are excluded. Free ribosomes also occur. Other cytoplasmic organelles include mitochondria, multivesicular bodies (MVB), microtubules, and glycogen granules (Gly).

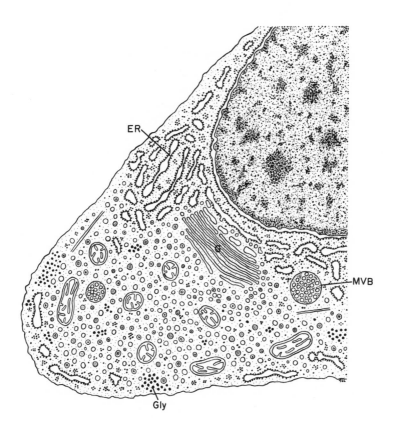

Fig. 17. Granule-containing cell of a planarian. The majority of the membrane-bounded granules in this cell type are about 800 A in diameter and composed of opaque material. Granules of lower density and small vesicles are also present. Several Golgi complexes (G) occur in the cell and may contain granules within their lamellae. Cisternae of rough-surfaced endoplasmic reticulum (ER) are present around the nucleus and peripheral cytoplasm. Several cisternae occur in a large stack, occupying a region of cytoplasm devoid of other organelles. Free ribosomes, mitochondria, and microtubules (MT) are present in the cytoplasm. N, nucleus.

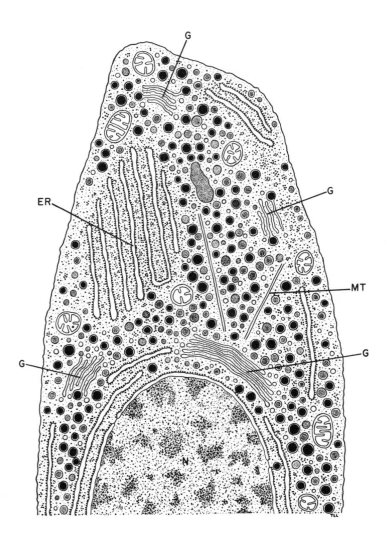

ally they contain a small, dense core. The cells also contain small amounts of endoplasmic reticulum, glycogen granules, free ribosomes, and multivesicular bodies. The vesicles closely resemble those in the nerve endings thought to be synaptic vesicles containing acetylcholine. Similar cells were not observed in *Procotyla*. Ganglion cells of *Procotyla* contain clear vesicles, especially in the Golgi zone, but the vesicles are not nearly so numerous as in the cells of *Dugesia*.

The other cell types contain granules and can be subdivided on the basis of size and density of the granules (Figs. 17–19). These cells occur on the periphery of the brain external to the ganglion cells and are also distributed throughout the nerve cords. They are not so numerous as ganglion cells and not so tightly packed, usually occurring singly. Granule-containing cells are oval or spindle-shaped and are more uniform in shape than ganglion cells. Processes may arise from each end of the cell, but profiles of some cells are devoid of processes.

The granule-containing cells are similar in cytoplasmic structure except for their granules. The nucleus is oval, centrally situated, and bounded by an envelope that contains pores. The nucleolus is not so large as in ganglion cells and may be absent. Chromatin material often occurs in small clumps in the nucleoplasm. Mitochondria, free ribosomes, rough-surfaced cisternae of endoplasmic reticulum, and a few lysosomes occur in the cytoplasm. In some cells a stack of rough-surfaced cisternae of endoplasmic reticulum is confined to a peripheral region of cytoplasm devoid of other organelles. Microtubules are present, although not nearly so numerous as in ganglion cells.

The Golgi apparatus of granule-containing cells is prominent. Some cells contain several Golgi complexes, most of which are arranged around the nucleus. Each organelle is composed of membranous lamellae and vesicles. Membrane-bounded granules surround the Golgi area and sometimes occur within the ends of dilated lamellae.

Three distinct morphological types of granules occur. (1) Granules 800 A in diameter with electron-opaque contents (Fig. 17); the homogeneous opaque material usually is separated from the membrane of the vesicle by a thin rim of low density. Less commonly, the opaque material occupies the entire vesicle. (2) Granules 800 A in diameter with contents of medium density (Fig. 18); the granule is the same size as the first but the contents are less dense and finely granular. The contents of these vesicles usually occupy a central core. (3) Granules 1,200 A in diameter with moderately dense contents (Fig. 19); the granule contains material of medium density but is larger than the others. Its contents are also separated from the membrane by a clear halo.

In each granule-containing cell one type of granule predominates. Usually, however, at least a few of each type of granule can be identified in every cell. Moreover, there is considerable variation in the structure of granules; many are intermediate in size and density from the three basic types. There is a continuous spectrum of size from 600 to 1,400 A and of density from medium to opaque. The three basic types described, however, are more numerous than the intermediate granules.

The granules that are 800 A in diameter resemble those observed in adrenergic cells and nerves of higher animals (Grillo and Palay, 1962; Richardson, 1962; Pellegrino de Iraldi et al., 1963). Epinephrine and norepinephrine are thought to occur in these granules (Pellegrino de Iraldi and DeRobertis, 1963; Van Orden et al., 1966). Coupland and Hopwood (1966) have shown that norepinephrine-containing granules are opaque after glutaraldehyde–osmium fixation because of formation of a glutaraldehyde–norepinephrine complex. Epinephrine granules are of medium density. Because glutaraldehyde was employed as a fixative in the studies on planaria described here, the 800-A opaque granules may contain norepinephrine, and the 800-A moderately dense granules could contain epinephrine. These cells, furthermore, are very similar to the catecholamine-containing cells in higher animals.

The 1,200-A granules are identical with neurosecretory granules

Fig. 18. Granule-containing cell of a planarian. The membrane-bounded granules in this cell are mostly 800 A in diameter and of medium density. A few granules have opaque contents and some have a density intermediate between medium and opaque. A halo of low density separates the contents from the enveloping membrane. The granules occupy most of the cytoplasmic space. A few cisternae of rough-surfaced endoplasmic reticulum, free ribosomes, and mitochondria are present. A Golgi apparatus (G) is situated near the nucleus and a centriole (C) occurs adjacent to this organelle.

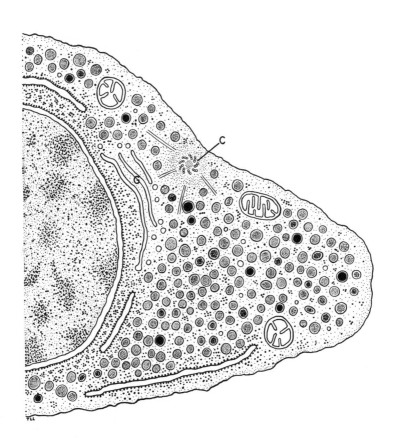

Fig. 19. Neurosecretory cell of a planarian. This cell is characterized by the presence of large membrane-bounded granules (NSG). Most of the granules are of medium density although a few are very dense. An elaborate Golgi apparatus occurs adjacent to the nucleus and contains dense material within the ends of its lamellae. Small vesicles with dense contents occur in the Golgi region. Free ribosomes (R) are abundant, and may occur in clusters of four to seven. The cytoplasm also contains cisternae of endoplasmic reticulum and mitochondria.

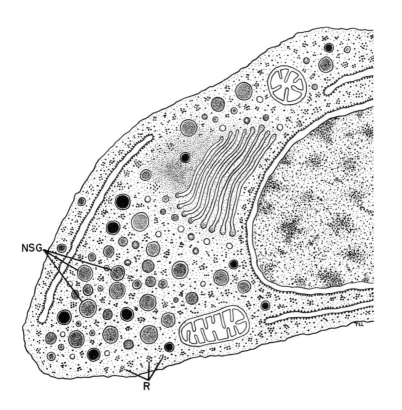

described in a large number of invertebrates and vertebrates (see Bern and Hagadorn, 1965). Neurosecretory substance has been identified by staining methods (Lender and Klein, 1961). Physiological evidence for neurosecretion in planaria is provided by the observation of changes in neurosecretory cells during regeneration (Lender and Klein, 1961), indicating a hormonal function. For these reasons, the existence of neurosecretory cells in planaria is reasonably well established.

The neuropil and peripheral nerve cords are composed of large numbers of individual nerve fibers (Fig. 20) which are not enclosed by myelin or sheath cells and are tightly packed. They differ in size and content, some large fibers appearing relatively empty and most small fibers containing large numbers of vesicles, granules, or organelles. Glycogen granules, mitochondria, microtubules, ribosomes, and cisternae of endoplasmic reticulum occur in fibers along with vesicles and granules. The latter are identical with those observed in the perikarya of neurons. All types of dense granules are numerous in the nerve fibers. As in the cell bodies, many granules are intermediate in appearance between the most common types. Some large, dilated vesicles are empty. Although there appears to be some tendency for granules of one type to occur in a single process, there is considerable intermixing. The small vesicles may occur with dense granules, and all types of granules may be intermixed. Even when more than one type of granule occurs within an ending, each type is often segregated into a separate region of axoplasm. The small vesicles, especially, occur in tight clusters, even in the presence of granules.

Intermixing of dense granules in the endings is not surprising because at least several of each type of granule occur in every granule-containing cell. The presence of granules in association with small vesicles is, however, surprising, since ganglion cells containing vesicles are devoid of granules. Some of the vesicles could originate from the granule-containing cells, because vesicles occur in the Golgi region of these cells; but ganglion cells outnumber and contain more processes than granule-containing cells. On this basis, more fibers containing only vesicles would be expected. It is possible that some vesicles originating in the Golgi region of ganglion cells progressively accumulate dense material as they move down the axon. Alternatively, clear vesicles in the nerve endings may be capable of selectively taking up catecholamines.

Synapses occur in the nervous system of planaria. In *Procotyla* they consist of an accumulation of vesicles or granules adjacent to the plasma membrane of the nerve fiber. Little or no thickening of pre- and postsynaptic membranes occurs. In *Dugesia,* however, there is thickening of membranes in synaptic regions (MacRae, 1963; Morita and Best, 1966). In all cases, the space between pre- and postsynaptic membranes is about 200 A. Microtubules extend into the endings and mitochondria are often present. In many respects these synapses resemble those of the mammalian nervous system.

Accessory cells similar or identical to neuroglia have been described in *Dugesia* at the fine structural level (Morita and Best, 1966). These cells are interspersed among the nerve elements in the brain; they have a low hyaloplasmic density and contain few organelles. Presumed glial processes extend between nerve fibers in the neuropil. The processes are clear, with few organelles.

Several studies have been performed on the fine structure of photoreceptors in the eye of planaria (Wolken, 1958; Press, 1959; Röhlich and Török, 1961; MacRae, 1964, 1966; Röhlich, 1966). The photoreceptor is a bipolar neurosensory cell (Fig. 21). The distal photoreceptive cell process extends into the eyecup, which is formed by pigment cells. The distal process arises from a cell body outside the eyecup. Proximally, an axon arises from the cell body and extends into the cerebral ganglion.

The distal surface of the cell (rhabdome), which is the photo-sensitive element, is thrown up into tightly packed villi or tubules which form the striated border. The dendrite contains vesicles and large vacuoles, mitochondria, and numerous microtubules. A nucleus with a nucleolus is present in the cell body. An elaborate Golgi apparatus, mitochondria, multivesicular bodies, micro-

Fig. 20. Neuropil of the planarian *Procotyla fluviatilis*. Nerve fibers are closely packed together in the central region of the brain. All types of vesicular elements are present: small (475 A) vesicles (V) with lucent contents, dense-core vesicles (800 A) (DV) with contents of either moderate or extreme density, and large (1,200 A) neurosecretory granules (NSG) of moderate density. Various combinations of vesicles and granules occur in individual nerve fibers. The small vesicles often occur in tight clusters. Synaptic relationships are indicated by terminal swellings of fibers or accumulation of vesicles or granules adjacent to the plasma membrane. Some nerve fibers or portions of fibers are devoid of vesicles or granules. Other structures occurring in the fibers are microtubules, mitochondria, and elements of endoplasmic reticulum. Glial elements could not be identified in this species.

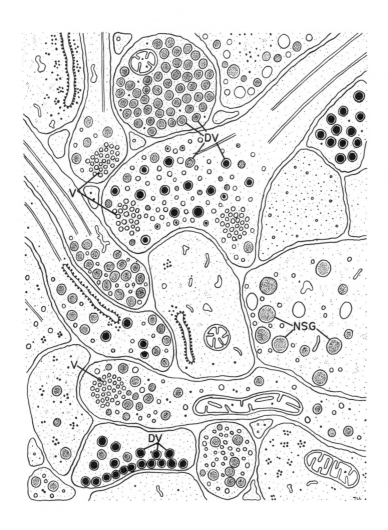

Fig. 21. Photoreceptor of the eye of a planarian. This cell is a bipolar sensory neuron. The distal process or dendrite extends into a pigmented eyecup (not illustrated) and terminates in an expanded retinal club. The surface membrane of the retinal club (rhabdome) is thrown up into many closely packed villi which form a striated border. The cytoplasm of the retinal club contains many mitochondria and microtubules and a few cisternae of endoplasmic reticulum. Microtubules (MT) are abundant in the dendrite. The cell body contains an oval nucleus with a nucleolus. Elongated cisternae of rough-surfaced endoplasmic reticulum are numerous in the perikaryon. A few mitochondria, ribosomes, multivesicular bodies, and Golgi complexes (G) occur. The basal process or axon extends from the cell body into the neuropil of the brain. Microtubules course longitudinally in the axon. The axon ends as a terminal swelling containing small vesicles (V), dense-core vesicles (DV), and mitochondria. A synaptic relationship is indicated by a thickening of the plasma membrane and accumulation of vesicles adjacent to the membrane.

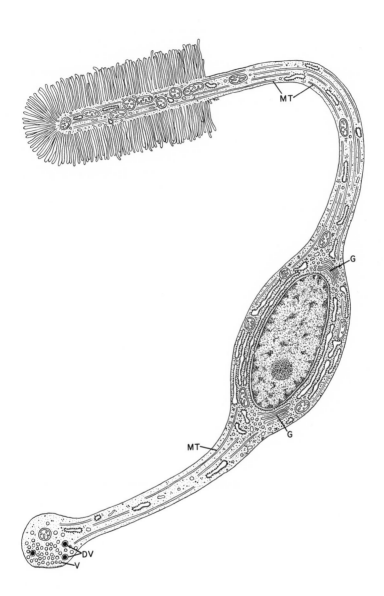

tubules, ribosomes, and elongated cisternae of rough-surfaced endoplasmic reticulum occur in the cytoplasm. The axon contains vesicles and microtubules. It extends into the neuropil of the brain and terminates as an enlargement which contains mitochondria, glycogen, clear synaptic vesicles, and larger dense vesicles. The terminations may form synapses with other neurites.

MacRae (1967) described sensory nerve endings within the auricular epithelium of *Dugesia*. The cells are similar to vertebrate olfactory cells, and it was suggested they could function as chemo-receptors. The cell bodies are below the epidermis and send distal fibers between epithelial cells. The nerve fibers contain micro-tubules, vesicles, and mitochondria and form club-shaped endings beyond the epithelial cells. One or two basal bodies and striated rootlets occur in the distal ending. A cilium extends outward from the basal body.

5

THE EVOLUTIONARY ORIGIN
OF THE NERVOUS SYSTEM

The origin of the primitive nerve cell

The nervous system probably arose in some multicellular form intermediate between the protozoa and lowest existing Eumetazoa, since even the latter have a nervous system that shows some specialization. The probable nature of the primitive ancestor of the Eumetazoa has been reconstructed (Chapter 1) to serve as a reference for the organization of the type of animal in which the nervous system may have appeared (Fig. 22). The ancestral form may have had a ciliated epidermis and an inner mass of undifferentiated cells and digestive cells. Effectors could have consisted of contractile processes in the epithelial cells, muscle cells, gland cells, pigment cells, and cilia. Lacking a nervous system, the behavior of the animal would have depended on the individual activities of its effectors and perhaps would have resembled that of some colonial protozoans. Although independent effectors alone may be suitable for small organisms, their uncoordinated activities would not permit complex behavior associated with a larger, multicellular animal.

The development of the nervous system would have high selective value, primarily because it would give the organism relatively greater independence of the environment. The animal could function as a unit; its behavior would not have to depend on the actions of individual and uncoordinated cells. It could obtain more precise information concerning the environment and respond accordingly. Depending on the nature of its secretory products, the nerve cell might also play a role in internal regulatory or homeostatic adjustments.

Fig. 22. Hypothetical reconstruction of the histology of the primitive nervous system as it may have appeared in a simple metazoan. The organism may have consisted of an outer epidermis and an inner solid mesenchyme. The epidermis is composed of ciliated epithelial cells (EC) some of which may have had muscular processes and thus have been epitheliomuscular cells (EmC). A few gland cells (GlC) are shown in the epidermis. The mesenchyme contains supportive or matrix-secreting mesenchymal cells (MsC), wandering phagocytic digestive cells (DC), gland cells (GlC), undifferentiated cells (UD), pigment cells (PC), and myocytes or muscle cells (MC). The nerve cells (black) form a loosely organized network. They are most abundant at the base of the epidermis, but a few occur in the mesenchyme. The loosely arranged cells contain two or more processes. The processes form no intimate connections with other cells, but some terminate in the vicinity of effectors (muscle, glands, epithelial cells, pigment cells) or near other neurons. Many processes come to a blind end in the intercellular spaces; others penetrate the epidermis.

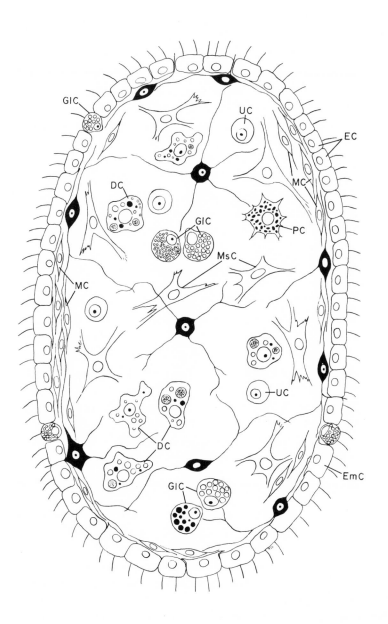

With a nervous system, the organism would be capable of a greater variety and complexity of behavior, expressed perhaps in improved feeding habits or ability to detect or escape from predators. If we assume that the nervous system would allow a greater variety of behavior, changes in behavior or habits might initiate shifts into new adaptive zones. The shifts in turn would lead to further evolutionary changes in the organism.

The nerve cell by definition transduces an environmental stimulus to an excited state, conducts the excitation, and transmits the information to effectors. The mechanisms by which nerve cells carry out these functions are widespread throughout the animal kingdom, suggesting the generality and great antiquity of the basic neurophysiological properties (Bullock, 1958). Excitability, conductivity, and elaboration of biologically active substances are properties of non-nervous cells and are present in the protozoa. Resting membrane potentials have been recorded in some protozoans (e.g. Kamada, 1934; Wolfson, 1943; Umrath, 1956; Hisada, 1957). There are receptors, such as the nonmotile cilia of some ciliates and the eyespots of flagellates, sensitive to specific stimuli. Graded (Umrath, 1956; Eckert, 1965a) and all-or-none responses (Hisada, 1957; Eckert, 1965a) have been observed. There are specialized effectors or effector regions separate from the receptors. In the dinoflagellate *Noctiluca miliaris,* mechanical stimulation evokes a graded potential which can give rise to an all-or-none action potential which is propagated from the source of the stimulus over the cell surface and triggers a bioluminescent flash (Eckert, 1965a,b). Although fibrillar structures have been described and have been thought to play a role in conduction and coordination of locomotor organelles in ciliates (Taylor, 1920), electrical properties of the cell membrane can explain these phenomena just as well (Bullock and Horridge, 1965a).

Protozoans contain neurohumors including acetylcholine (Bayer and Wense, 1936a; Bülbring et al., 1949) and catecholamines (Bayer and Wense, 1936b; Östlund, 1954; Janakidevi et al., 1966). Acetylcholinesterase has also been identified (Bayer and Wense, 1936a; Seamen and Houlihan, 1951). Acetylcholine and acetylcholinesterase may be related to the activity of ciliated and flagellated cells (Seaman and Houlihan, 1951; Willmer, 1960).

The nerve cell, being specialized and complicated, must have developed gradually over a long period of time as the result of a large number of small gene mutations. It is difficult, however, to determine the actual sequence of changes that occurred. Neurons are usually considered to be of epidermal origin, on the basis of the embryological development of nervous tissue from ectoderm and the close proximity of the outer layer of cells to the environment. Nerve cells occur in the epidermis of coelenterates and flatworms. It is possible, however, that they could have sunk internally in animals with a mesenchyme.

The prenervous cell could have been excitable and conductile to the extent that many non-nervous cells and protozoans are. If the prenervous cell was epidermal, it could mean that the cellular machinery for secretion, found to some extent in most epithelial cells, was present. An excitable and conductile cell with secretory capabilities might evolve into a neuron by specialization of these basic cytoplasmic characteristics. It is apparent from the nature of the evolved cell that the prenervous cell must have become sensitive to stimuli, responding by active depolarization. Mechanisms for selective permeability and active transport, therefore, appeared. The conductive properties of the cell became better developed by the appearance of elongated cytoplasmic processes and changes in the permeability characteristics of this portion of the plasma membrane. Finally, transmission had to take place between the evolving neuron and effectors.

As will be discussed later, there is evidence that transmission initially was accomplished by release of biologically active agents. The primitive nerve cell could have acquired the ability to produce these substances in two ways. First, it has been seen that neurohumors occur in protozoa, indicating that these substances could have evolved before the appearance of the neuron. Thus it is possible that they occurred in some cells of the primitive metazoan but subserved some other function. By a change in function, they could have been utilized during evolution of the nerve cell as transmitter agents. The alternative explanation is that the appearance of new enzyme systems enabled the ancestral cell, which already had the basic machinery for secretion, to produce these substances. In either case it was necessary that release of the substances be coupled with excitation and conduction and furthermore that the effectors become sensitive to the chemical agents. Thus it is suggested that the nerve cell arose by the coupling of electrical activity with secretion of biologically active substances so that a chain of events in response to stimuli resulted in alteration of effector activity.

Effectors are usually thought to have evolved before the nervous system (Pantin, 1956). As pointed out, however, the development of the nerve cell probably occurred over a long period, and it is likely that effectors such as fibrillar proteins capable of shortening were undergoing evolution and change at the same time the nerve cell was evolving.

Although the characteristics that have been outlined probably appeared during evolution of the nerve cell, it is not possible to determine their sequence of appearance or the nature of the intermediate stages. Grundfest (1965) has suggested that the ancestral cell was a secretory cell that developed specialized receptive surfaces, and that a conductile component appeared between the receptive and secretory regions to form a neuron. Although this hypothesis cannot be ruled out, it has no advantage over some of the other possibilities such as an excitable and conductive cell developing secretory capacity. It may be that the attempt to organize a stepwise progressive sequence with one function added on to another is artificial. It seems more reasonable to assume that the neuronal functions evolved simultaneously and were not formed in an assembly-line or sequential manner.

During differentiation of nerve cells from interstitial cells in hydra, the formation of flagella, cytoplasmic processes, and the organelles associated with secretion (endoplasmic reticulum, Golgi apparatus, vesicles) occurs simultaneously (Lentz, 1965a). During development, therefore, the receptive, conductile, and secretory portions of the cell are formed at the same time. Similarly, the nerve cell may have arisen when the basic functions, all evolving at the same time, reached a certain degree of specialization and operated in an interdependent fashion.

The nature of the primitive nerve cell

The structure and functional properties of the primitive nervous system, as they may have existed in a primitive multicellular animal ancestral to the Metazoa, will be described in this section. These suggestions are based on the evidence obtained from study of the existing primitive nervous systems and on the functions of evolved neurons that are considered to be simple or primitive. The primitive nerve cell in the metazoan ancestor had to have certain characteristics in order to be classified as a neuron: it had to respond by excitation to stimuli, conduct the excitation, and affect the activities of effectors.

In excitable cells, environmental changes or stimuli produce an energy-requiring response or active depolarization that is accompanied by a change in the membrane permeability to sodium or to sodium and potassium. Presumably, a portion of the plasma membrane of the primitive nerve cell was excitable. It should be noted that specialized receptors (e.g. cilia) are not necessary for response to stimuli. In sea anemones, for example, certain patterns of behavior in response to mechanical (Passano and Pantin, 1955) and light (Ross, 1966) stimulation are better explained by direct stimulation of nerves in the nerve net than on specific sensory receptors. The nerve cell initially may have been responsive to a wide range of chemical and physical stimuli, specialized receptors sensitive to specific stimuli developing later. Although sensory cells are often considered to be the first nerve cell type to have appeared (Parker, 1919; Ariëns Kappers, 1929), there is no evidence for the existence of a nervous system composed exclusively of sensory (or sensorimotor) cells. If the primitive nerve cell was sensitive to a wide range of stimuli, it could respond equally well to environmental stimuli or chemical agents released by other neurons. In this case it would be capable of sensory, internuncial, and/or motor function.

The earliest type of response is considered to be graded (Bishop, 1956; Grundfest, 1959; Bullock and Horridge, 1965a); its intensity depends on the strength of the stimulus and it is conducted decrementally. The graded response is seemingly more primitive and generalized than the all-or-none response (Bishop, 1956). Many non-nervous cells exhibit graded responses and, in neurons, examples of graded activity occur in sensory terminals, dendrites, and efferent axon terminals. In lower animals, graded potentials have been described in protozoa, neurons of sea anemones after mechanical stimulation (Passano and Pantin, 1955), and planarian photoreceptors after light stimulation (Behrens, 1961). Graded, local conducting systems have been observed in some colonial polyps (Josephson, 1961).

After initiation of graded responses in nerve processes, conduction could take place in the neuron but, being decremental, it would be effective only over relatively short distances. The local response, however, may sometimes spread for long distances, especially if it is moving into more excitable regions of tissue (Hodgkin, 1938). In crustaceans, local graded potentials are capable of initiating muscular contraction (Katz, 1949). Bishop (1956) has pointed out that if the neuron is small it can perform its essential functions by means of graded responses without the intervention of all-or-none impulses.* Pantin (1950) has also suggested that local graded potentials spread over a neuronal network might be sufficient for coordination in very small animals. When a relatively great distance separated receptors from effectors, as in large animals, the graded response and decremental

---

*It should be emphasized that the graded response is not a passive depolarization but is an active process involving a change in the permeability of the cell membrane. It is not necessarily spread electrotonically but can be propagated actively, atlhough decrementally, over short distances.

conduction would not be an effective means of conduction. Thus, according to this theory, the all-or-none response developed later, allowing conduction of excitation over distances too great to be covered by a decremental graded response (Bishop, 1956).

Most of the evidence suggests that transmission occurred by release of biologically active substances rather than by electrotonic coupling. First, in hydra, neurons have unspecialized nerve endings often terminating in intercellular spaces. Passive depolarization or electrotonic transmission could not accomplish transmission between cells separated by relatively large spaces. Electrotonic transmission has been correlated with synapses characterized by close apposition or fusion of cell membranes (Bennett et al., 1963; Dewey and Barr, 1964), but this type of morphological specialization has not been observed in the nervous systems of primitive organisms. Finally, as discussed later, chemosensitivity and electrical inexcitabilty of effectors appear to be more primitive than electrical excitability (Bullock and Horridge, 1965a).

Welsh (1955, 1959) has defined the categories of chemical agents produced by nerve cells. Neurohumors, also known as chemical mediators or transmitter agents, are chemical agents released by neurons, acting over short distances on neurons, muscles, and glands, and usually being enzymatically destroyed. Neurosecretory substances are longer acting, more stable, and act over greater distances. All these substances have a coordinating or hormonal function. Neurohormone is an inclusive term for the regulatory agents, both neurohumors and neurosecretory substances, produced by nerve cells.

In certain cases the substances classified as neurohumors, especially catecholamines, may act diffusely over relatively large distances and are not rapidly destroyed. In these situations they can be regarded as hormones and their cells of origin neurosecretory cells (Scharrer, 1967). It seems preferable, however, to retain a distinction between these cells and the cells usually regarded as neurosecretory which produce polypeptide hormones.

111

Neurohormones are present in primitive nervous systems. Acetylcholine, epinephrine, norepinephrine, and 5-hydroxytryptamine occur in protozoa and invertebrate nervous systems (Table 1). Neurosecretory substances have been identified in some cells of sponges and in nerve cells of hydra and planaria. It seems possible that because these substances are of general occurrence in nervous systems, including those of the lowest animals, they could have been present in the primitive nervous system. The secretion of neurohormones may have been a modification of the basic secretory capacity of the prenervous cells. Coupling of the excitatory and conductile properties of the cell with secretion allowed transmission to occur after stimuli.

Table 1. Neurohormones Identified in Lower Organisms

Protozoa
  Acetylcholine
    Trypanosoma rhodesiense (Bülbring et al., 1949)
    Paramecium (Bayer and Wense, 1936a)
  Acetylcholinesterase
    Paramecium (Bayer and Wense, 1936a)
    Tetrahymena geleii (Seaman and Houlihan, 1951)
  Catecholamines
    Noctiluca miliaris (Östlund, 1954)
    Crithidia fasciculata (Janakidevi et al., 1966)
    Paramecium (Bayer and Wense, 1936b)
    Tetrahymena pyriformis (Janakidevi et al., 1966)
Porifera
  Acetylcholinesterase
    Sycon ciliatum (Lentz, 1966b)
  Catecholamines
    Sycon ciliatum (Lentz, 1966b)
  5-Hydroxytryptamine
    Sycon ciliatum (Lentz, 1966b)
  Neurosecretory substance
    Sycon ciliatum (Lentz, 1966b)
    Tethya lyncurium (Pavans de Ceccatty, 1966a)
Coelenterata
  Acetylcholine
    Aurelia aurita (Mitropolitanskaya, 1941)
    Actinia equina (Mitropolitanskaya, 1941)

Acetylcholinesterase
Tubularia crocea (Bullock and Nachmansohn, 1942)
Hydra fusca (Mitropolitanskaya, 1941)
Hydra littoralis (Lentz and Barrnett, 1961)
Aurelia aurita (Bullock and Nachmansohn, 1942)
Cyanea capillata (Bullock and Nachmansohn, 1942)
Actinia equina (Mitropolitanskaya, 1941)
Metridium marginatum (Bullock and Nachmansohn, 1942)
Sagartia luciae (Bullock and Nachmansohn, 1942)
Sagartia parasitica (Augustinsson, 1948)
Catecholamines
Hydra littoralis (Wood and Lentz, 1964)
Rhizostoma pulmo (Östlund, 1954)
Alcyonium digitatum (Östlund, 1954)
Metridium senile (Dahl et al., 1963; Wood and Lentz, 1964)
Metridium dianthus (Östlund, 1954)
Tealia felina (Dahl et al., 1963)
5-Hydroxytryptamine
Hydra littoralis (Wood and Lentz, 1964)
Hydra oligactis (Welsh and Moorhead, 1960)
Calliactis parasitica (Mathias et al., 1960)
Metridium senile (Welsh and Moorhead, 1960)
Sagartia luciae (Welsh and Moorhead, 1960)
Neurosecretory substance
Hydra littoralis (Lentz and Barrnett, 1965a)
Hydra pseudoligactis (Burnett et al., 1964)
Platyhelminthes
Acetylcholine
Planaria maculata (Welsh, 1946)
Acetylcholinesterase
Planaria maculata (Bullock and Nachmansohn, 1942)
Procotyla fluviatilis (Bullock and Nachmansohn, 1942)
Catecholamines
Turbellaria (Dahl et al., 1963)
5-Hydroxytryptamine
Dugesia dorotocephala (Welsh and Moorhead, 1960)
Dugesia tigrina (Welsh and Moorhead, 1960)
Neurosecretory substance
Dugesia gonocephala (Oosaki and Ishii, 1965)
Dugesia dorotocephala (Morita and Best, 1965, 1966)
Polycelis nigra (Lender and Klein, 1961)
Procotyla fluviatilis (Lentz, 1967b)

113

The suggestion that transmission was accomplished by means of neurohumors gains support if the effectors were highly sensitive to neurohumors. Striated and nonstriated muscle probably appeared very early in evolution (Prosser, 1967). The muscle tissue in sponges, hydra, and planaria can be classified as smooth, with two types of myofilaments (MacRae, 1963; Bagby, 1966; Lentz, 1966a). Striated muscle with a banding pattern similar to that in vertebrates occurs in coelenterates (Chapman et al., 1962). Smooth muscle cells are sensitive over their entire surface to neural transmitters. Myocytes of sponges do not respond to electrical stimulation (Prosser, 1967). Similarly, striated muscle in its primitive state presumably had a high and uniform sensitivity over the entire cell to transmitters such as acetylcholine (Ginetsinskii and Shamarina, 1942, *in* Thesleff, 1961; Loewi, 1945; Thesleff, 1961; Grundfest, 1965). In the muscle fibers of higher animals, chemosensitivity is restricted to the motor end plate (see Thesleff, 1961), and this region is electrically inexcitable (Grundfest, 1961, 1965; Werman, 1963). After denervation the entire muscle membrane becomes highly sensitive to transmitter substances. Thus the restriction of chemosensitivity to sites of synaptic contact may be the result of innervation. Prior to this stage, however, the entire muscle cell was chemosensitive and electrically inexcitable.

It is usually assumed that neurosecretory cells originated later and were derived from ordinary nerve cells (Hanström, 1954; Barrington, 1964; Gabe, 1966). It has also been suggested that neurosecretory cells might be derived from epidermal secretory cells that acquired neural characteristics (Clark, 1956a,b). Neurosecretory cells certainly must have appeared early, since they are present in coelenterates and flatworms.

If hormonal substances were present in the ancestral metazoan, they probably occurred in the nervous system. Scharrer and Scharrer (1963) have pointed out that, progressing higher on the evolutionary scale, the proportion of the nervous system that produces hormones becomes smaller, this function being increasingly relegated to glands outside the nervous system. In invertebrates, practically all the hormones controlling bodily functions such as reproduction, metabolism, growth, and behavior are produced by neurosecretory cells. The Scharrers suggest that the nervous system was originally the sole agency of nervous and endocrine functions. Thus, whatever hormones the primitive organism contained, they were most likely formed by the primitive nerve cells comprising the nervous system. Nothing is known, however, of the possible nature of the hormonal substances. Unlike the small nonpeptide neurohumors, the polypeptide hormones have undergone considerable evolution and change.

The possibility that neurohumors and neurosecretory substance both may have been present in the primitive nerve cell should not be surprising. Nerve cells and neurosecretory cells are basically similar (Scharrer and Scharrer, 1954; Koelle, 1959; Bern, 1966), and the secretory products of neurons can exist in various combinations in individual cells. Neurosecretory cells are capable of responding to stimuli such as changes in temperature and osmotic pressure and to several chemical agents. They conduct action potentials (Potter and Loewenstein, 1955; Cross and Green, 1959; Morita et al., 1961; Bennett and Fox, 1962) and contain synaptic vesicles identical with those that contain acetylcholine in other nerve cells (DeRobertis, 1962). Some ordinary neurons may be capable of secreting hormonal substances in addition to transmitters (Lentz, 1967c). Evidence has been accumulating, also, that more than one type of transmitter can occur in a nerve ending. Small, clear synaptic vesicles thought to contain acetylcholine occur in the presence of larger, dense core vesicles that may contain catecholamines. These observations indicate that the classification of nerve cells into types on the basis of their products is not so rigid as previously supposed. In fact, combinations of humoral and hormonal substances may be of common occurrence and could represent the primitive situation.

Neurohumors, as well as neurosecretory substances, are capable of acting over distances greater than the width of the synaptic cleft. For example, circulating catecholamines produced in the adrenal medulla have an effect on peripheral tissues. Some smooth muscle cells may be activated by general diffusion of transmitter in contrast to release of transmitter from nerve endings at discrete sites on the surface of the muscle (Rosenblueth, 1950; Burnstock and Holman, 1961). In smooth muscle there appears to be no postsynaptic specialization like the end plate of skeletal muscle. Along with the generalized surface sensitivity of primitive muscle to transmitters, these examples indicate that humoral agents released by the primitive nerve cell could have influenced effectors situated at a relatively large distance from the nerve cell. Because diffusion of hormonal substances is effective, specialized synapses may not have been present in the primitive nervous system. Processes of primitive nerve cells may have terminated in intercellular spaces in the vicinity of, but not necessarily on, the effector cells. Synapses probably appeared later as morphological specializations of the nerve ending with the effector, allowing more precise, rapid, and polarized transmission. Enzymatic mechanisms for the destruction of neurohumors may have been present, but not necessarily, because diffusion, desensitization of receptors, or binding of the transmitter can also accomplish inactivation (Gerschenfeld, 1966).

In hydra the nervous system may be associated more with slower long-term events than with rapid responses. Nerve endings are unspecialized and, although action potentials have been detected, these may originate and be conducted by epitheliomuscular cells (Josephson and Macklin, 1967). On the other hand, neurohumors and neurosecretory substances have been identified in the nerve cells. One function of the neurohumors may be to alter the threshold of nematocyst discharge. Although nematocysts respond directly to environmental mechanical and chemical stimuli, discharge is facilitated by the presence of neurohumors (Lentz and Barrnett, 1962). Neurosecretory substances of hydra and planaria appear to be related to control of growth and development (Lender and Klein, 1961; Lentz, 1965b,c). Sponges are not capable of rapid responses and conductivity is limited. If the cells in sponges containing neurohormones have any effect on other cells, these influences must be slow acting and diffuse. For these reasons, primitive nerve cells probably served at first simply to modulate the activity of the effectors. The effectors themselves could respond to stimuli and could be classified as independent effectors although, like nematocysts, they were influenced by nervous activity. In this case the effects of the primitive nervous system may have been to alter the excitability, spontaneous activity, or resting membrane potential of the effectors. The role of neurosecretory substances in growth and regeneration in lower invertebrates furthermore suggests that trophic effects may have been an important function of the primitive nervous system.

Evolution of the nerve cell

The primitive nerve cell had all the basic properties required by a nerve cell, but these existed in relatively simplified form. By subsequent specialization of one or another basic function and the appearance of new features, the primitive nerve cell could give rise to the different types of neurons. If the cell developed an apical specialization and became highly sensitive to specific external stimuli and its processes terminated on another neuron, it could be considered a sensory cell in the usual sense. Internuncial and motor neurons became specialized for conducting impulses between sensory cells and effectors over long distances. It might be expected that these cells developed an all-or-none response, that their transmitter agents were restricted to a small number of neurohumors released at nerve endings, and that specialized synapses occurred between cells.

I consider the nervous system of the primitive ancestral organism to have been composed of a single basic nerve cell type which was capable of sensory, internuncial, or motor function. In hydra, in addition to sensory and motor types, cells with sensorimotor or internuncial-motor functions occur. The latter may represent intermediate stages in the evolution of the generalized cell into those with only sensory, internuncial, or motor function. It is possible that deeply situated ganglion and neurosecretory cells with internuncial-motor functions might also be sensitive to external stimuli, as are some cells in the nerve net of anemones (Passano and Pantin, 1955; Ross, 1966). In this case, these cells may closely resemble the primitive nerve cell.

The primitive nerve cell may have been capable of producing neurohumors and neurosecretory substance and thus could have given rise to both ordinary neurons and neurosecretory cells. Initially, both classes of substances may have acted hormonally. Ordinary neurons became specialized principally for the formation of neurohumors and release of the substances at synaptic contacts on effectors so that these agents act as neurotransmitters. By becoming specialized primarily for the secretion of polypeptide hormonal substances and release of these substances into intercellular spaces or near vascular systems, the primitive nerve cell would have given rise to neurosecretory cells. In higher animals, cases where the neurohumors act over large distances and thus have a hormonal function (pp. 111, 115), or where combinations of neurohumors and neurosecretory substance occur in the same cell (p. 115) may indicate that some cells retain or have the ability to revert to the primitive condition.

Sponges may represent a stage at or slightly before the stage in the evolution of the nervous system when the hypothetical primitive nerve cell arose. Cells are present in sponges that contain neurohumors and possibly neurosecretory substance. Some of these cells and those considered by Pavans de Ceccatty to be nerves have long processes extending to the surface of the animal, between each other, and ending near effectors such as myocytes or in intercellular spaces. Pavans de Ceccatty (1966a,b) also showed that specialized connections exist between some cells. Sponges also respond to external stimuli and are capable of slow conduction. What is not known is whether the neurohumors are released after excitation and have some influence on effectors. Although some characteristics of nerve cells are present in sponges, there is no evidence that the neural features are so combined in the cells that they significantly affect the behavior of the organism in response to external stimuli. It is likely that a situation similar to that in sponges preceded the appearance of the nervous system.

Hydra shows a considerable advance over sponges. The nervous system of hydra contains some features of more evolved systems but in other respects is relatively undifferentiated. At least three distinct nerve cell types (sensory, ganglion, and neurosecretory) are present. These cells, however, may represent different stages of the same cell type and are continuously being renewed by loss of aged cells and replacement by differentiation from interstitial cells. They contain cytological specializations characteristic of higher nerve cells, including vesicles, dense granules, extensive Golgi apparatus, and microtubules (neurotubules). On the other hand, sheath cells and morphologically specialized synapses are absent and the processes are of one type. Neurohumors and neurosecretory substance are present in this primitive nervous system. The nervous system of hydra may be one of the simplest existing systems in that it can be definitely identified as a system, and yet it contains fewer specialized characteristics than other systems. For this reason it appears that study of this animal might be most likely to reveal the primitive or elemental properties of nervous tissue.

There are certain similarities in the structure of the nerve cell types in hydra. Secretory and ganglion cells differ only by the absence of dense membrane-bounded granules in the latter. Sensory cells possess a cilium or apical specialization but otherwise resemble ganglion cells. Neurosensory cells are deeply situated ganglion or secretory cells that have a cilium extending to the surface. The basic similarity of these cells supports the notion of a common origin.

Neurohumors including acetylcholine, epinephrine, norepinephrine, and 5-hydroxytryptamine occur in hydra neurons. Neurosecretory substance with a hormonogenic function is also present. In higher nervous systems, catecholamines and neurosecretory substance occur in dense membrane-bounded granules which can usually be distinguished from each other. In hydra, however, there appears to be only one type of granule-containing cell that could contain both catecholamines and neurosecretory substance.

The nerve cells of hydra are arranged into a configuration known as a nerve net. As pointed out by Bullock and Horridge (1965a), the concept of a nerve net does not necessarily imply continuity of neurons but refers to the diffuseness and availability of many conduction pathways. The nerve net of hydra appears to be composed of separate neurons. The nerve cell terminations, however, do not contain the morphological specializations of synapses of higher animals. Although the presence of granules or vesicles in the nerve endings suggests at least a polarization of function for the individual cell, it appears that transmission can occur in either direction across these unspecialized nerve terminations. This situation correlates well with the properties of the nerve net, including diffuseness of conduction and high relative autonomy.

Impulses may arise in and be conducted by epitheliomuscular cells (Josephson and Macklin, 1967). Conduction in this layer of cells may account for some of the properties attributed to the nerve net. It would be of interest to determine the relationship of the nervous system to the epithelial conducting system.

In other coelenterates, electrical excitability and all-or-none impulses have been detected (Horridge, 1954; Josephson, 1964). Synapses morphologically similar to those of higher animals have been observed (Horridge and Mackay, 1962; Jha and Mackie, 1967). Sense organs occur, and the beginnings of centralization are suggested by the accumulation of neurons into ganglia.

The nervous system reaches a higher degree of complexity in the planarians. A central nervous system is present, and the brain is necessary for some of the activities of the animal. The brain plays a role in regeneration (Lender and Klein, 1961) and possibly in the acquisition of conditioned responses (Thompson and McConnell, 1955). The nerve cords mediate the activity of the brain and local reflexes.

In coelenterates, local concentrations of cells occur to form ganglia, but this system is too diffuse to be considered central.

In primitive acoels, the nervous system consists of an epidermal network best developed anteriorly and with some suggestion of longitudinal strands. The anterior concentration of nerve cells and nerve cords is situated at the base of the epidermis and is connected with the epidermal plexus by numerous fine branches. In more complicated turbellarians, the cephalic ganglia (brain) and nerve cords have sunk internally into the mesenchyme. This sequence suggests that the central nervous system arose from the diffuse nerve network. If the nerve plexus became thickened anteriorly and within longitudinal strands, it could give rise to the brain and longitudinal cords respectively.

At the cellular level there is a greater variety of nerve cells. Ganglion cells, which resemble those of hydra, are most abundant in the cephalic ganglia. In hydra there is only one type of granule-containing cell to account for catecholamines and neurosecretory substance. In planaria, separate cells exist with granules morphologically similar to those containing epinephrine, norepinephrine, and neurosecretory substance in higher animals. Several types of sensory cells are present and occur singly or organized into complicated sense organs. Finally, morphologically polarized synapses are present.

Evolution of photoreceptors appears to have occurred along two lines (Eakin, 1963). In the first type of photoreceptor, disks or flattened sacs arise in close relationship to a cilium. This receptor, exemplified by the vertebrate rods and cones, is found in coelenterates, echinoderms, protochordates, and vertebrates. The second type of photoreceptor is an array of microvilli or tubules, called a rhabdomere, not developmentally associated with a ciliary apparatus. This type occurs in the flatworms, annelids, mollusks, and arthropods.

The major building blocks of nervous tissue are present in the flatworms. Nerve cells with the cytologic features of the major classes of neurons are present. Specialized synapses occur between neurons. Other types of cells associated with neurons in higher animals, such as specialized glia and Schwann cells capable of forming myelin sheaths, are absent or not prominent. Besides the appearance of these cells, subsequent evolution of the nervous system largely involves reorganization of the architectural arrangement of the basic cell types.

Summary

A hypothesis has been proposed to explain the origin of the nervous system and the structural and functional characteristics of the primitive nerve cell. These suggestions are based on observations of existing primitive nervous systems and on the functions of evolved systems that are thought to be basic or primitive. I believe the nerve cell arose gradually over a period of time, appearing first in a primitive multicellular animal ancestral to the present Eumetazoa. Evolution of the nerve cell may have involved the development of a chain of electrical events related to secretion so that effector activity was marshalled. The essential characteristics that had to be present in the primitive nerve cell, as well as in any nerve cell, are excitability in response to stimuli, conduction of the excitation, and transmission of the information to effectors. The prenervous cell may have been excitable and conductile to the extent that protozoans and many non-nervous cells are. It may also have contained the cellular machinery for secretion that is characteristic of many epithelial cells. By specialization of these cytoplasmic characteristics, a cell of this type could give rise to a nerve cell. The basic functions evolved simultaneously and were coupled so that they operated in a coordinated sequence characteristic of the neuron.

The primitive nervous system is visualized as a loosely organized group of cells and their processes in a simple free-living form with effectors that may have included muscle cells, gland cells, and cilia. Environmental stimuli impinging on the organism resulted in a graded response in the nerve cell that could be conducted over short distances. Depolarization was followed by release of biologically active substances. Effectors, such as muscle, may have been electrically inexcitable and chemosensitive over their entire surface in their primitive state. These cells would be affected by neurohumors released by the primitive nerve cell. The neurons, however, probably served only to modulate or alter the activity of the effectors, which still retained the capacity to respond directly to environmental stimuli. Another important function of the primitive nervous system may have been trophic effects, related perhaps to control of growth or development, mediated by hormonal secretory substances. As both neurohumors and neurosecretory substances are capable of acting over relatively large distances, specialized synapses may not have been present in the primitive nervous system. In this case, the functions of the primitive nerve cell were largely hormonal and it most closely resembled a neurosecretory cell.

Other nerve cell types arose from the primitive nerve cell by specialization of certain of its basic functions or by the appearance of new characteristics. Sensory cells evolved through specialization of the receptive surface, and other cells acquired predominantly internuncial and motor functions. As the organism became larger and its nerve cell processes longer, the all-or-none response developed, allowing conduction over longer distances. Specialized synapses appeared, permitting rapid and polarized transmission. Some cells became specialized for the production and release of neurohumors that functioned as neurotransmitters, while others became principally neurosecretory. The central nervous system probably arose through localized thickening and accumulation of neurons in the diffuse nerve network.

Many early and intermediate stages in the evolution of the nervous system appear to be present in the Porifera, Coelenterata, and Platyhelminthes. However, most of the structural and functional features of the neurons of the highest animals, as well as the basic nerve cell types, can also be found in these phyla. The differences in the behavior of these lower forms and higher invertebrates or vertebrates are obvious. Thus, evolution of the nervous system beyond these primitive animals involved either the appearance of some as yet undefined characteristics or, more likely, rearrangement of the structural relationships of the basic cellular units.

REFERENCES

Ankel, W. E., and G. Wintermann-Kilian, 1952. Eine bei *Ephydatia fluviatilis* neu gefundene hochdifferenzierte Zellart und die Struktur der Doppelepithelien. *Z. Naturforsch., 7b:* 475–81.

Applewhite, P. B., and H. J. Morowitz, 1966. The micrometazoa as model systems for studying the physiology of memory. *Yale J. Biol. Med., 39:* 90–105.

Ariëns Kappers, C. U., 1929. *The Evolution of the Nervous System in Invertebrates, Vertebrates, and Man.* Bohn, Haarlem, 335 pp.

Augustinsson, K. B., 1948. Cholinesterases. A study in comparative enzymology. *Acta Physiol. Scand., 15,* suppl. 52: 1–182.

Bacq, Z. M., 1935. Recherches sur la physiologie et la pharmacologie du système nerveux autonome. XVI. Les esters de la choline dans les extraits de tissus des invertébrés. *Arch. Int. Physiol., 42:* 24–42.

———— 1947. L'acétylcholine et l'adrénaline chez les invertébrés. *Biol. Rev., 22:* 73–91.

Bagby, R. M., 1966. The fine structure of myocytes in the sponges *Microciona prolifera* (Ellis and Solander) and *Tedania ignis* (Duchassaing and Michelotti). *J. Morphol., 118:* 167–82.

Barrington, E. J. W., 1964. *Hormones and Evolution.* Van Nostrand, Princeton, 154 pp.

Bayer, G., and T. Wense, 1936a. Über den Nachweis von Hormonen in einzelligen Tieren. I. Cholin und Acetylcholin im *Paramecium. Pflüg. Arch. ges. Physiol., 237:* 417–22.

———— 1936b. Über den Nachweis von Hormonen in einzelligen Tieren. II. Adrenalin (Sympathin) im *Paramecium. Pflüg. Arch. ges. Physiol., 237:* 651–54.

Behrens, M. C., 1961. The electrical response of the planarian photoreceptor. *Comp. Biochem. Physiol., 5:* 129–38.

Bennett, M. V. L., and S. Fox, 1962. Electrophysiology of caudal neurosecretory cells in the skate and fluke. *Gen. Comp. Endocrinol., 2:* 77–95.

References

Bennett, M. V. L., E. Aljure, Y. Nakajima, and G. D. Pappas, 1963. Electrotonic junctions between teleost spinal neurons: Electrophysiology and ultrastructure. *Science, 141:* 262–64.

Bern, H. A., 1966. On the production of hormones by neurons and the role of neurosecretion in neuroendocrine mechanisms. *Symp. Soc. Exp. Biol., 20:* 325–44.

Bern, H. A., and I. R. Hagadorn, 1965. Neurosecretion, in: T. H. Bullock and G. A. Horridge, *Structure and Function in the Nervous Systems of Invertebrates,* vol. 1. Freeman, San Francisco, pp. 356–432.

Bishop, G. H., 1956. Natural history of the nerve impulse. *Physiol. Rev., 36:* 376–99.

Brien, P., 1960. The fresh water hydra. *Am. Scientist, 48:* 461–75.

Brien, P., and M. Reniers-Decoen, 1949. La croissance, la blastogenèse, l'ovogenèse chez *Hydra fusca* (Pallas). *Bull. Biol. France Belg., 82:* 293–386.

Bülbring, E., E. M. Lourie, and U. Pardoe, 1949. The presence of acetylcholine in *Trypanosoma rhodesiense* and its absence from *Plasmodium gallinaceum. Brit. J. Pharmacol., 4:* 290–94.

Bullock, T. H., 1958. Evolution of neurophysiological mechanisms, in: A. Roe and G. G. Simpson, eds., *Behavior and Evolution.* Yale Univ. Press, New Haven, 165–77.

Bullock, T. H., and G. A. Horridge, 1965a. *Structure and Function in the Nervous Systems of Invertebrates,* vol. 1. Freeman, San Francisco, 798 pp.

——— 1965b. Ibid., vol. 2, pp. 809–1719.

Bullock, T. H., and D. Nachmansohn, 1942. Choline esterase in primitive nervous systems. *J. Cell Comp. Physiol., 20:* 239–42.

Burnett, A. L., and N. A. Diehl, 1964a. The nervous system of hydra. I. Types, distribution and origin of nerve elements. *J. Exp. Zool., 157:* 217–26.

——— 1964b. The nervous system of hydra. III. The initiation of sexuality with special reference to the nervous system. *J. Exp. Zool., 157:* 237–50.

Burnett, A. L., T. Lentz, and M. Warren, 1960. The nematocyst of hydra. I. The question of control of the nematocyst discharge reaction by fully fed hydra. *Ann. Soc. Roy. Zool. Belg., 90:* 247–68.

Burnett, A. L., R. Davidson, and P. Wiernik, 1963. On the presence of a feeding hormone in the nematocyst of *Hydra pirardi. Biol. Bull., 125:* 226–33.

Burnett, A. L., N. A. Diehl, and F. Diehl, 1964. The nervous system

126

of hydra. II. Control of growth and regeneration by neurosecretory cells. *J. Exp. Zool., 157:* 227–36.

Burnstock, G., and M. E. Holman, 1961. The transmission of excitation from autonomic nerve to smooth muscle. *J. Physiol., Lond., 155:* 115–33.

Chapman, D. M., C. F. A. Pantin, and E. A. Robson, 1962. Muscle in coelenterates. *Rev. Canad. Biol., 21:* 267–78.

Child, C. M., 1904a. Studies on regulation. V. The relation between the central nervous system and regeneration in *Leptoplana:* Posterior regeneration. *J. Exp. Zool., 1:* 463–512.

———— 1904b. Studies on regulation. VI. The relation between the central nervous system and regulation in *Leptoplana:* Anterior and lateral regeneration. *J. Exp. Zool., 1:* 513–58.

———— 1910. The central nervous system as a factor in the regeneration of polyclad Turbellaria. *Biol. Bull., 19:* 333–38.

Chun, C., 1880. Die Ctenophoren des Golfes von Neapel. *Fauna Flora Neapel,* Monogr. 1, 313 pp.

———— 1881. Die Natur und Wirkungswiese der Nesselzellen bei Coelenterates. *Zool. Anz., 4:* 646–50.

Clark, R. B., 1956a. On the transformation of neurosecretory cells into ordinary nerve cells. *Fysiogr. Sällsk. Lund, Förh., 26:* 82–89.

———— 1956b. On the origin of neurosecretory cells, *Ann. Sci. Nat. Zool.* (11) *18:* 199–207.

Claus, C., 1878. Studien über Polypen und Quallen der Adria. *Denkschr. Akad. Wiss. Wien, 38:* 1–64.

Clayton, D. E., 1932. A comparative study of the non-nervous elements in the nervous systems of invertebrates. *J. Entomol. Zool., 24:* 3–22.

Corning, W. C., and E. R. John, 1961. Effect of ribonuclease on retention of conditioned response in regenerated planarians. *Science, 134:* 1363–65.

Coupland, R. E., and D. Hopwood, 1966. The mechanism of the differential staining reaction for adrenalin- and noradrenalin-storing granules in tissues fixed in glutaraldehyde. *J. Anat., Lond., 100:* 227–43.

Cross, B. A., and J. D. Green, 1959. Activity of single neurons in the hypothalamus: Effect of osmotic and other stimuli. *J. Physiol., Lond., 148:* 554–69.

Dahl, E., B. Falck, C. von Mecklenburg, and H. Myhrberg, 1963. An adrenergic nervous system in sea anemones. *Quart. J. Micr. Sci., 104:* 531–34.

References

Davenport. D., D. M. Ross, and L. Sutton, 1961. The remote control of nematocyst discharge in the attachment of *Calliactis parasitica* to shells of hermit crabs. *Vie et Milieu, 12:* 197–209.

deBeer, G. R., 1963. The evolution of Metazoa, in: J. Huxley, A. C. Hardy, and E. B. Ford, eds., *Evolution As a Process.* Collier, New York, pp. 34–45.

DeRobertis, E., 1962. Ultrastructure and function in some neurosecretory systems, in: H. Heller and R. B. Clark, eds., *Neurosecretion.* Academic Press, New York, pp. 3–20.

DeRobertis, E., G. R. de Lores Arnaiz, L. Salganicoff, A. Pellegrino de Iraldi, and L. M. Zieher, 1963. Isolation of synaptic vesicles and structural organization of the acetylcholine system within brain nerve endings. *J. Neurochem., 10:* 225–35.

Dewey, M. M., and L. Barr, 1964. A study of the structure and distribution of the nexus. *J. Cell Biol., 23:* 553–86.

Eakin, R. M., 1963. Lines of evolution of photoreceptors, in: D. Mazia and A. Tyler, eds., *General Physiology of Cell Specialization,* McGraw-Hill, New York, pp. 393–425.

Eakin, R. M., and J. A. Westfall, 1962. Fine structure of photoreceptors in the hydromedusan *Polyorchis penicillatus. Proc. Nat. Acad. Sci., U. S., 48:* 826–33.

Eckert, R., 1965a. Bioelectric control of bioluminescence in the dinoflagellate *Noctiluca.* I. Specific nature of triggering events. *Science, 147:* 1140–42.

———— 1965b. II. Asynchronous flash initiation by a propagated triggering potential. *Science, 147:* 1142–45.

Emson, R. H., 1966. The reactions of the sponge *Cliona celata* to applied stimuli. *Comp. Biochem. Physiol., 18:* 805–27.

Ewer, R. F., 1947. On the functions and mode of action of the nematocysts of hydra. *Proc. Zool. Soc. Lond., 117:* 365–76.

Florey, E., 1962. Comparative neurochemistry: Inorganic ions, amino acids and possible transmitter substances of invertebrates, in: K. A. C. Elliott, I. H. Page, and J. H. Quastel, eds., *Neurochemistry.* Thomas, Springfield, Ill., pp. 673–93.

Forrest, H., 1962. Lack of dependence of the feeding reaction in hydra on reduced glutathione. *Biol. Bull., 122:* 343–62.

Gabe, M., 1966. *Neurosecretion,* Pergamon Press, Oxford, 872 pp.

Gelei, J. von, 1930. "Echte" freie Nervenendigungen (Bemerkungen zu den Receptoren der Turbellarian). *Z. Morph. Ökol. Tiere, 18:* 786–98.

Gerschenfeld, H. M., 1966. Chemical transmitters in invertebrate nervous systems. *Symp. Soc. Exp. Biol., 20:* 299–324.

References

Gibbins, J. R., L. G. Tilney, and K. R. Porter, 1966. Microtubules in primary mesenchyme cells of sea urchin embryos. *Anat. Rec., 154:* 347.

Graff, L. von, 1912–17. Turbellaria, Tricladida, in: *H. G. Bronn's Klassen und Ordnungen des Tier-Reichs,* Winter'sche, Leipzig, pp. 2601–3369.

Grillo, M. A., and S. L. Palay, 1962. Granule containing vesicles in the autonomic nervous system, vol. 2, *Fifth Intl. Cong. Electron Micr.* Academic Press, New York, U–1.

Grundfest, H., 1959. Evolution of conduction in the nervous system, in: A. D. Bass, ed., *Evolution of Nervous Control from Primitive Organisms to Man.* Amer. Assoc. Adv. Sci., Publ. 52, Washington, pp. 43–86.

———— 1961. Ionic mechanisms in electrogenesis. *Ann. N. Y. Acad. Sci., 94:* 405–57.

———— 1965. Evolution of electrophysiological properties among sensory receptor systems, in: J. W. S. Pringle, ed., *Essays on Physiological Evolution.* Pergamon Press, Oxford, 107–38.

Hadzi, J., 1909. Über das Nervensystem von Hydra. *Arb. Zool. Inst. Univ. Wien, 17:* 225–68.

———— 1953. An attempt to reconstruct the system of animal classification. *Syst. Zool., 2:* 145–54.

———— 1963. *The Evolution of the Metazoa.* Pergamon Press (Macmillan), New York, 499 pp.

Haeckel, E., 1874. Die Gastraea-Theorie, die phylogenetische Classification des Thierreichs und die Homologie der Keimblätter. *Jena Z. Naturwiss., 8:* 1–55.

Hanson, E. D., 1958. On the origin of the Eumetazoa. *Syst. Zool., 7:* 16–47.

Hanström, B., 1954. On the transformation of ordinary nerve cells into neurosecretory cells. *Fysiogr. Sällsk. Lund, Förh., 24:* 75–82.

Haug, G. 1933. Die Lichtreaktionen der Hydren. *Z. vergl. Physiol., 19:* 246–303.

Havet, J., 1901. Contribution a l'étude du système nerveux des Actinies. *La Cellule, 18:* 385–419.

Hertwig, O., and R. Hertwig, 1878. *Das Nervensystem und die Sinnesorgane der Medusan.* Vogel, Leipzig, 186 pp.

———— 1879–80. Die Actinien anatomisch und histologisch mit besonderer Berücksichtigung des Nervenmuskelsystems untersucht. *Jena Z. Naturwiss.,* N.F. 6: 475–640; 7: 39–89.

Hisada, M., 1957. Membrane resting and action potentials from a protozoan, *Noctiluca scintillans. J. Cell Comp. Physiol., 50:* 57–71.

References

Hodgkin, A. L., 1938. The subthreshold potentials in a crustacean nerve fibre. *Proc. Roy. Soc. Lond., B, 126:* 87–121.

Horridge, G. A., 1954. The nerves and muscles of Medusae. I. Conduction in the nervous system of *Aurellia aurita* Lamarck. *J. Exp. Biol., 31:* 594–600.

Horridge, G. A., and B. Mackay, 1962. Naked axons and symmetrical synapses in coelenterates. *Quart. J. Micr. Sci. 103:* 531–41.

Hovey, H. B., 1929. Associative hysteresis in marine flatworms. *Physiol. Zool., 2:* 322–33.

Hyman, L. H., 1940. The Invertebrates, vol. 1: *Protozoa Through Ctenophora.* McGraw-Hill, New York, 726 pp.

———— 1942. The transition from the unicellular to the multicellular individual. *Biol. Symp., 8:* 27–42.

———— 1951. The Invertebrates, vol. 2: *Platyhelminthes and Rhynchocoela.* McGraw-Hill, New York, 550 pp.

Jacobson, A. L., 1963. Learning in flatworms and annelids. *Psychol. Bull., 60:* 74–94.

———— 1965. Learning in planarians: Current status. *Animal Behavior,* suppl. 1: 76–81.

Janakidevi, K., V. C. Dewey, and G. W. Kidder, 1966. The biosynthesis of catecholamines in two genera of protozoa. *J. Biol. Chem., 241:* 2576–78.

Jha, R. K., and G. O. Mackie, 1967. The recognition, distribution and ultrastructure of hydrozoan nerve elements. *J. Morphol., 123:* 43–62.

Jones, C. S., 1947. The control and discharge of nematocysts in hydra. *J. Exp. Zool., 105:* 25–61.

Jones, W. C., 1957. The contractility and healing behavior of pieces of *Leucosolenia complicata. Quart. J. Micr. Sci., 98:* 203–17.

———— 1962. Is there a nervous system in sponges? *Biol. Rev., 37:* 1–50.

Josephson, R. K., 1961. Colonial responses of hydroid polyps. *J. Exp. Biol., 38:* 559–77.

————1964. Coelenterate conducting systems, in: R. F. Reiss, ed., *Neural Theory and Modeling.* Stanford Univ. Press, Stanford, pp. 414–22.

———— 1965. Mechanisms of pacemaker and effector integration in coelenterates. *Symp. Soc. Exp. Biol., 20:* 33–47.

Josephson, R. K., and M. Macklin, 1967. Transepithelial potentials in *Hydra. Science, 156:* 1629–31.

Kamada, T., 1934. Some observations on potential differences across the ectoplasm membrane of *Paramecium. J. Exp. Biol., 11:* 94–102.

Katz, B., 1949. Neuro-muscular transmission in invertebrates. *Biol. Rev., 24:* 1–20.

Kleinenberg, N., 1872. *Hydra. Eine anatomisch-entwicklungsgeschichtliche Untersuchungen.* Engleman, Leipzig, 90 pp.

Koehler, O., 1932. Sinnesphysiologie der Süsswasserplanarien. *Z. vergl. Physiol., 16:* 606–756.

Koelle, G. B., 1959. Neurohumoral agents as a mechanism of nervous integration, in: A. D. Bass, ed., *Evolution of Nervous Control from Primitive Organisms to Man.* Am. Assoc. Adv. Sci., Publ. 52, Washington, pp. 87–114.

Ledbetter, M. C., and K. R. Porter, 1963. A "microtubule" in plant cell fine structure. *J. Cell Biol., 19:* 239–50.

Lendenfeld, R. von, 1885a. The histology and nervous system of the calcareous sponges. *Proc. Linn. Soc. N. S. Wales, 9:* 977–83.

———— 1885b. Das Nervensystem der Spongien. *Zool. Anz., 8:* 47–50.

———— 1885c. A nervous system in sponges. *Am. Naturalist, 19:* 611.

———— 1886. Contributions toward the knowledge of the nervous and muscular systems of the horny sponges. *Ann. Mag. Nat. Hist., 17:* 372–77.

———— 1887. The function of nettle-cells. *J. Roy. Micr. Soc., 7:* 247–48.

———— 1889. *Horny Sponges.* Trübner, London, 936 pp.

———— 1892. Die Spongien der Adria. I. Die Kalkschwämme. *Z. wiss. Zool., 53:* 185–321, 361–433.

Lender, T., 1955. Mise en évidence et propriétés de l'organisme de la régénération des yeux de la planaire *Polycelis nigra. Rev. Suisse Zool., 62:* 268–75.

Lender, T., and N. Klein, 1961. Mise en évidence de cellules sécrétrices dans le cerveau de la Planaire *Polycelis nigra.* Variation de leur nombre au cours de la régénération posterieure. *C. R. Acad. Sci., Paris, 253:* 331–33.

Lenhoff, H. M., 1961. Activation of the feeding reflex in *Hydra littoralis.* I. Role played by reduced glutathione, and quantitative assay of the feeding reflex. *J. Gen. Physiol., 45:* 331–34.

Lentz, T. L., 1965a. The fine structure of differentiating interstitial cells in *Hydra. Z. Zellforsch., 67:* 547–60.

———— 1965b. Fine structural changes in the nervous system of the regenerating hydra. *J. Exp. Zool., 159:* 181–94.

———— 1965c. Hydra: Induction of supernumerary heads by isolated neurosecretory granules. *Science, 150:* 633–35.

———— 1966a. *The Cell Biology of Hydra.* North-Holland, Amsterdam; Interscience (Wiley), New York, 199 pp.

131

———— 1966b. Histochemical localization of neurohumors in a sponge. *J. Exp. Zool., 162:* 171–80.

———— 1967a. Rhabdite formation in planaria: The role of microtubules. *J. Ultrastruct. Res., 17:* 114–26.

———— 1967b. Fine structure of nerve cells in a planarian. *J. Morphol., 121:* 323–38.

———— 1967c. Fine structure of nerves in the regenerating limb of the newt *Triturus. Am. J. Anat., 121:* 647–670.

Lentz, T. L., and R. J. Barrnett, 1961. Enzyme histochemistry of hydra. *J. Exp. Zool., 147:* 125–49.

———— 1962. The effect of enzyme substrates and pharmacological agents on nematocyst discharge. *J. Exp. Zool., 149:* 33–38.

———— 1963. The role of the nervous system in regenerating hydra: The effect of neuropharmacological agents. *J. Exp. Zool., 154:* 305–28.

———— 1965a. Fine structure of the nervous system of *Hydra. Am. Zool., 5:* 341–56.

———— 1965b. Surface specializations of hydra cells: The effect of enzyme inhibitors on ferritin uptake. *J. Ultrastruct. Res., 13:* 192–211.

Lesh, G. E., and A. L. Burnett, 1966. An analysis of the chemical control of polarized form in hydra. *J. Exp. Zool., 163:* 55–78.

Loewi, O., 1945. Aspects on the transmission of the nervous impulse. II. Theoretical and clinical implications. *J. Mt. Sinai Hosp., 12:* 851–65.

Loomis, W. F., 1955. Glutathione control of the specific feeding reactions of hydra. *Ann. N. Y. Acad. Sci., 62:* 209–28.

Mackie, G. O., 1965. Conduction in the nerve-free epithelia of siphonophores. *Am. Zool., 5:* 439–53.

MacRae, E. K., 1963. Observations on the fine structure of pharyngeal muscle in the planarian *Dugesia tigrina. J. Cell Biol., 18:* 651–62.

———— 1964. Observations on the fine structure of photoreceptor cells in the planarian *Dugesia tigrina. J. Ultrastruct. Res., 10:* 334–49.

———— 1966. The fine structure of photoreceptors in a marine flatworm. *Z. Zellforsch., 75:* 469–84.

———— 1967. Fine structure of sensory nerve endings within planarian auricular epithelium. *Anat. Rec., 157:* 282.

Mariscal, R. N., 1966. The symbiosis between tropical sea anemones and fishes: A review, in: R. I. Bowman, ed., *The Galápagos.* Univ. California Press, Berkeley, pp. 157–71.

Mathias, A. P., D. M. Ross, and M. Schachter, 1960. The distribution of 5-hydroxytryptamine, tetramethylammonium, homarine, and other substances in sea anemones. *J. Physiol., Lond., 151:* 296–311.

McConnell, C. H., 1932. The development of the ectodermal nerve net in the buds of hydra. *Quart. J. Micr. Sci., 75:* 495–509.

McConnell, J. V., 1965. Cannibals, chemicals, and contiguity. *Animal Behavior,* suppl. 1: 61–66.

McConnell, J. V., A. L. Jacobson, and D. P. Kimble, 1959. The effects of regeneration upon retention of a conditioned response in the planarian. *J. Comp. Physiol. Psychol., 52:* 1–5.

McConnell, J. V., R. Jacobson, and B. M. Humphries, 1961. The effects of ingestion of conditioned planaria on the response level of naive planaria: A pilot study. *Worm Run. Dig., 3:* 41–47.

McCullough, C. B., 1965. Pacemaker interaction in *Hydra. Am. Zool., 5:* 499–504.

McNair, G. T., 1923. Motor reactions of the fresh-water sponge, *Ephydatia fluviatilis. Biol. Bull., 44:* 153–66.

Metschnikoff, E., 1879. Spongiolische Studien. III. Entwicklungsgeschichtliches über die Kalkschwämme. *Z. wiss. Zool., 32:* 362–71.

Mitropolitanskaya, R. L., 1941. On the presence of acethylcholin and cholinestherase in the protozoa, spongia, and coelenterata. *C. R. (Doklady) Acad. Sci. URSS, 31:* 717–18.

Morita, H., T. Ishibashi, and S. Yamashita, 1961. Synaptic transmission in neurosecretory cells. *Nature, 191:* 183.

Morita M., and J. B. Best, 1965. Electron microscopic studies on planaria. II. Fine structure of the neurosecretory system in the planarian *Dugesia dorotocephala. J. Ultrastruct. Res., 13:* 396–408.

——— 1966. Electron microscopic studies of planaria. III. Some observations on the fine structure of planarian nervous tissue. *J. Exp. Zool., 161:* 391–412.

Müller, H. G., 1936. Untersuchungen über spezifische Organe niederer Sinne bei rhabdocoelen Turbellarien. *Z. vergl. Physiol., 23:* 253–92.

Murbach, L., 1893. Zur Entwicklung der Nesselorgane bei den Hydroiden. *Zool. Anz., 16:* 174–75.

Olmsted, J. M. D., 1922. The role of the nervous system in the regeneration of polyclad Turbellaria. *J. Exp. Zool., 36:* 48–56.

Oosaki, T., and S. Ishii, 1965. Observations on the ultrastructure of nerve cells in the brain of the planarian, *Dugesia gonocephala. Z. Zellforsch., 66:* 782–93.

Orton, J. H., 1924. An experimental effect of light on the sponge, *Oscarella. Nature, 113:* 924–25.

Östlund, E., 1954. The distribution of catechol amines in lower animals and their effect on the heart. *Acta Physiol. Scand., 31,* suppl. 112: 1–67.

Pantin, C. F. A., 1942. The excitation of nematocysts. *J. Exp. Biol., 19:* 294–310.

———— 1950. Behavior patterns in lower invertebrates. *Symp. Soc. Exp. Biol., 4:* 175–95.

———— 1952. The elementary nervous system. *Proc. Roy. Soc. Lond., B, 140:* 147–68.

———— 1956. The origin of the nervous system. *Pubbl. Staz. Zool. Napoli, 28:* 171–81.

———— 1965. Capabilities of the coelenterate behavior machine. *Am. Zool., 5:* 581–89.

Parker, G. H., 1910. The reactions of sponges, with a consideration of the origin of the nervous system. *J. Exp. Zool., 8:* 1–41.

———— 1918. Some underlying principles in the structure of the nervous system. *Science, 47:* 151–62.

———— 1919. *The Elementary Nervous System,* Lippincott, Philadelphia, 229 pp.

Passano, L. M., 1963. Primitive nervous systems. *Proc. Nat. Acad. Sci., U. S., 50:* 306–13.

Passano, L. M., and C. B. McCullough, 1962. The light response and the rhythmic potentials of *Hydra. Proc. Nat. Acad. Sci., U. S., 48:* 1376–82.

———— 1964. Co-ordinating systems and behavior in *Hydra.* I. Pacemaker system of the periodic contractions. *J. Exp. Biol., 41:* 643–64.

———— 1965. Co-ordinating systems and behavior in *Hydra.* II. The rhythmic potential system. *J. Exp. Biol., 42:* 205–31.

Passano, L. M., and C. F. A. Pantin, 1955. Mechanical stimulation in the sea-anemone *Calliactis parasitica. Proc. Roy. Soc. Lond., B, 143:* 226–38.

Pavans de Ceccatty, M., 1955. Le système nerveux des Éponges calcaires et siliceuses. *Ann. Sci. Nat. Zool.* (11) *17:* 203–90.

———— 1960. Les structures cellulaires de type nerveux et de type musculaire de l'Éponge siliceuse *Tethya lyncurium* Lmk. *C. R. Acad. Sci., Paris, 251:* 1818–19.

———— 1966a. Ultrastructures et rapports des cellules mésenchymateuses de type nerveux de l'Éponge *Tethya lyncurium* Lmk. *Ann. Sci. Nat. Zool.* (12) 8: 577–614.

———— 1966b. Connexions cellulaires et jonctions polarisées du réseau intramésenchymateux, chez l'Éponge *Hippospongia communis* Lmk. *C. R. Acad. Sci., Paris, 263:* 145–47.

References

Pavans de Ceccatty, M., M. Gargouil, and E. Coraboeuf, 1960. Les réactions motrices de l'Éponge *Tethya lyncurium* à quelques stimulations expérimentales. *Vie et Milieu, 11:* 594–600.

Pedersen, K. J., 1964. The cellular organization of *Convoluta convoluta,* an acoel turbellarian: A cytological, histochemical and fine structural study. *Z. Zellforsch., 64:* 655–87.

Pellegrino de Iraldi, A., and E. DeRobertis, 1963. Action of reserpine, iproniazid and pyrogallol on nerve endings of the pineal gland. *Int. J. Neuropharm., 2:* 231–39.

Pellegrino de Iraldi, A., H. F. Duggan, and E. DeRobertis, 1963. Adrenergic synaptic vesicles in the anterior hypothalamus of the rat. *Anat. Rec., 145:* 521–31.

Picken, L. E. R., and R. J. Skaer, 1966. A review of researches on nematocysts. *Symp. Zool. Soc. Lond., 16:* 19–50.

Potter, D. D., and W. R. Loewenstein, 1955. Electrical activity of neurosecretory cells. *Am. J. Physiol., 183:* 652.

Press, N., 1959. Electron microscope study of the distal portion of a planarian retinular cell. *Biol. Bull., 117:* 511–17.

Prosser, C. L., 1967. Problems in comparative physiology of nonstriated muscles, in: C. A. G. Wiersma, ed., *Invertebrate Nervous Systems.* Univ. Chicago Press, pp. 133–49.

Prosser, C. L., T. Nagai, and R. A. Nystrom, 1962. Oscular contractions in sponges. *Comp. Biochem. Physiol., 6:* 69–74.

Richardson, K. C., 1962. The fine structure of autonomic nerve endings in smooth muscle of the rat vas deferens. *J. Anat., Lond., 96:* 427–42.

Robertson, J. A., 1928. Reaction of *Polycelis* in relation to physiological polarity. *Biol. Zbl., 48:* 427–30.

Robson, E. A., 1965. Adaptive changes in Cnidaria. *Animal Behavior,* suppl. 1: 54–59.

Röhlich, P., 1966. Sensitivity of regenerating and degenerating planarian photoreceptors to osmium fixation. *Z. Zellforsch., 73:* 165–73.

Röhlich, P., and L. T. Török, 1961. Elektronenmikroskopische Untersuchungen des Auges von Planarien. *Z. Zellforsch., 54:* 362–81.

Rosenblueth, A., 1950. *The Transmission of Nerve Impulses at Neuroeffector Junctions and Peripheral Synapses.* Technology Press of Mass. Inst. of Technology, and John Wiley, New York, 325 pp.

Ross, D. M., 1965. The behavior of sessile coelenterates in relation to some conditioning experiments. *Animal Behavior,* suppl. 1: 43–52.

———— 1966. The receptors of the Cnidaria and their excitation. *Symp. Zool. Soc. Lond., 16:* 413–18.

Rushforth, N. B., 1965. Behavioral studies of the coelenterate *Hydra pirardi* Brien. *Animal Behavior,* suppl. 1: 30–42.

Rushforth, N. B., A. L. Burnett, and R. Maynard, 1963. Behavior in hydra: Contraction responses of *Hydra pirardi* to mechanical and light stimuli. *Science, 139:* 760–61.

Scharrer, B., 1967. The neurosecretory neuron in neuroendocrine regulatory mechanisms. *Am. Zool., 7:* 161–69.

Scharrer, E., and B. Scharrer, 1954. Hormones produced by neurosecretory cells. *Recent Prog. Hormone Res., 10:* 183–240.

———— 1963. *Neuroendocrinology.* Columbia Univ. Press, New York, 289 pp.

Seaman, G. R., and R. K. Houlihan, 1951. Enzyme systems in *Tetrahymena geleii* S. II. Acetylcholinesterase activity. Its relation to motility of the organism and to coordinated ciliary action in general. *J. Cell Comp. Physiol., 37:* 309–21.

Semal-Van Gansen, P., 1952. Note sur le système nerveux de l' hydra. *Acad. Roy. Belg., Bull. Classe Sci., 38:* 718–35.

Singer, R. H., N. B. Rushforth, and A. L. Burnett, 1963. The photodynamic action of light on hydra. *J. Exp. Zool., 154:* 169–74.

Sollas, W. J., 1888. The Tetractinellida. *Challenger Reports, 21:* 1–513.

Spangenberg, D. B., and R. G. Ham, 1960. The epidermal nerve net of hydra. *J. Exp. Zool., 143:* 195–202.

Stewart, C., 1885. In: F. J. Bell, *Comparative Anatomy and Physiology.* Lea, Philadelphia, p. 431.

Taliaferro, W. H., 1920. Reactions to light in *Planaria maculata,* with special reference to the function and structure of the eyes. *J. Exp. Zool., 31:* 59–116.

Taylor, C. V., 1920. Demonstration of the function of the neuromotor apparatus in *Euplotes* by the method of microdissection. *Univ. Calif. Publ. Zool., 19:* 403–70.

Thesleff, S., 1961. Nervous control of chemosensitivity in muscle. *Ann. N. Y. Acad. Sci., 94:* 535–46.

Thompson, R., and J. McConnell, 1955. Classical conditioning in the planarian, *Dugesia dorotocephala. J. Comp. Physiol. Psychol., 48:* 65–68.

Tilney, L. G., and K. R. Porter, 1965. Studies on microtubules in Heliozoa. I. The fine structure of *Actinosphaerium nucleofilum* (Barrett) with particular reference to the axial rod structure. *Protoplasma, 60:* 317–44.

References

Trembley, A., 1744. *Mémoires pour servir à l'histoire d'un genre de polypes d'eau douce, a brâs en forme de cornes.* Verbeek, Leyden, 324 pp.

Tuzet, O., and M. Pavans de Ceccatty, 1953a. Les cellules nerveuses de *Grantia compressa pennigera* Haeckel (Éponge calcaire Hétêro-coele). *C. R. Acad. Sci., Paris, 235:* 1541–43.

———— 1953b. Les cellules nerveuses de l'Éponge calcaire homocoele *Leucandra johnstonni* Cart. *C. R. Acad. Sci., Paris, 236:* 130–33.

———— 1953c. Les cellules nerveuses et neuro-musculaires de l'Éponge *Cliona celata* Grant. *C. R. Acad. Sci., Paris, 236:* 2342–44.

———— 1953d. Les lophocytes de l'Éponge *Pachymatisma johnstonni* Bow. *C. R. Acad. Sci., Paris, 237:* 1447–49.

———— 1953e. Les cellules nerveuses de l'Éponge *Pachymatisma johnstonni* Bow. *C. R. Acad. Sci., Paris, 237:* 1559–61.

Tuzet, O., R. Loubatières, and M. Pavans de Ceccatty, 1952. Les cellules nerveuses de l'Éponge *Sycon raphanus* O. S. *C. R. Acad. Sci., Paris, 234:* 1394–96.

Umrath, K., 1956. Elektrische Messungen und Reizversuche an *Amoeba proteus. Protoplasma, 47:* 347–58.

Van Orden, L. S., F. E. Bloom, R. J. Barrnett, and N. J. Giarman, 1966. Histochemical and functional relationships of catecholamines in adrenergic nerve endings. I. Participation of granular vesicles. *J. Pharmacol. Exp. Ther., 154:* 185–99.

Welsh, J. H., 1946. Evidence of a trophic action of acetylcholine in a planarian. *Anat. Rec., 94:* 421.

———— 1955. Neurohormones, in: G. Pincus and K. V. Thimann, eds., *The Hormones,* vol. 3. Academic Press, New York, pp. 97–151.

———— 1959. Neuroendocrine substances, in: A. Gorbman, ed., *Comparative Endocrinology.* Wiley, New York, pp. 121–33.

Welsh, J. H., and M. Moorhead, 1960. The quantitative distribution of 5-hydroxytryptamine in the invertebrates especially in their nervous systems. *J. Neurochem., 6:* 146–69.

Werman, R., 1963. Electrical inexcitability of the frog neuro-muscular synapse. *J. Gen. Physiol., 46:* 517–31.

Westblad, E., 1937. Die Turbellarien-Gattung *Nemertoderma* Stein-böck. *Acta Soc. Fauna Flora fenn., 60:* 45–89.

Whittaker, V. P., I. A. Michaelson, and R. J. A. Kirtland, 1964. The separation of synaptic vesicles from nerve ending particles ('Synaptosomes'). *Biochem. J., 90:* 293–303.

References

Willmer, E. N., 1960. *Cytology and Evolution*. Academic Press, New York, 430 pp.

Wilson, E. B., 1891. The heliotropism of *Hydra*. *Am. Naturalist, 25:* 414–33.

Wintermann, G., 1951. Entwicklungsphysiologische Untersuchungen an Süsswasserschwämmen. *Zool. Jb. (Anat.), 71:* 427–86.

Wolfson, C., 1943. Potential-difference measurements on *Chaos chaos. Physiol. Zool., 16:* 93–100.

Wolken, J. J., 1958. Studies of photoreceptor structures. *Ann. N. Y. Acad. Sci., 74:* 164–81.

Wood, J. G., and T. L. Lentz, 1964. Histochemical localization of amines in *Hydra* and in the sea anemone. *Nature, 201:* 88–90.

Wood, R. L., 1959. Intercellular attachment in the epithelium of hydra as revealed by electron microscopy. *J. Biophys. Biochem. Cytol., 6:* 343–52.

Zelman, A., L. Kabat, R. Jacobson, and J. V. McConnell, 1963. Transfer of training through injection of "conditioned" RNA into untrained planarians. *Worm Run. Dig., 5:* 14–21.

AUTHOR INDEX

Aljure, E., 111, 126
Ankel, W. E., 22, 125
Applewhite, P. B., 43, 125
Ariëns Kappers, C. U., 109, 125
Augustinsson, K. B., 53, 113, 125

Bacq, Z. M., 26, 53, 125
Bagby, R. M., 15, 114, 125
Barr, L., 111, 128
Barrington, E. J. W., 114, 125
Barrnett, R. J., 42, 44, 45, 52, 56, 91, 113, 116, 132, 137
Bayer, G., 106, 112, 125
Behrens, M. C., 110, 125
Bennett, M. V. L., 111, 115, 125, 126
Bern, H. A., 96, 115, 126
Best, J. B., 78, 84, 97, 113, 133
Bishop, G. H., 110, 111, 126
Bloom, F. E., 91, 137
Brien, P., 44, 126
Bülbring, E., 106, 112, 126
Bullock, T. H., 11, 26, 42, 53, 71, 75, 78, 84, 106, 110, 111, 113, 120, 126
Burnett, A. L., 40, 45, 48, 49, 52, 53, 61, 64, 113, 126, 132, 136
Burnstock, G., 116, 127

Chapman, D. M., 114, 127
Child, C. M., 74, 127
Chun, C., 5, 44, 127
Clark, R. B., 10, 114, 127
Claus, C., 5, 127
Clayton, D. E., 78, 127
Coraboeuf, E., 18, 19, 135
Corning, W. C., 74, 127
Coupland, R. E., 91, 127
Cross, B. A., 115, 127

Dahl, E., 56, 84, 113, 127
Davenport, D., 45, 128
Davidson, R., 40, 126
deBeer, G. R., 4, 128
deLores Arnaiz, G. R., 60, 128
DeRobertis, E., 60, 91, 115, 128, 135
Dewey, M. M., 111, 128
Dewey, V. C., 106, 112, 130
Diehl, F., 53, 64, 113, 126
Diehl, N. A., 49, 52, 53, 64, 113, 126
Duggan, H. F., 91, 135

Eakin, R. M., 61, 68, 121, 128
Eckert, R., 106, 128
Emson, R. H., 19, 128
Ewer, R. F., 40, 128

Falck, B., 56, 84, 113, 127
Florey, E., 26, 128
Forrest, H., 40, 128
Fox, S., 115, 125

Gabe, M., 53, 114, 128
Gargouil, M., 18, 19, 135
Gelei, J. von, 78, 128
Gerschenfeld, H. M., 116, 128
Giarman, N. J., 91, 137
Gibbins, J. R., 60, 129
Ginetsinskii, A. G., 114
Graff, L. von, 78, 129
Green, J. D., 115, 127
Grillo, M. A., 91, 129
Grundfest, H., 10, 108, 110, 114, 129

Hadzi, J., 4, 49, 129
Haeckel, E., 3, 129
Hagadorn, I. R., 96, 126

Ham, R. G., 49, 136
Hanson, E. D., 3, 129
Hanström, B., 114, 129
Haug, G., 40, 129
Havet, J., 7, 129
Hertwig, O., 5, 6, 7, 129
Hertwig, R., 5, 6, 7, 129
Hisada, M., 106, 129
Hodgkin, A. L., 110, 130
Holman, M. E., 116, 127
Hopwood, D., 91, 127
Horridge, G. A., 8, 11, 42, 68, 71, 75, 78, 106, 110, 111, 120, 126, 130
Houlihan, R. K., 107, 112, 136
Hovey, H. B., 71, 130
Humphries, B. M., 74, 133
Hyman, L. H., 3, 45, 130

Ishibashi, T., 115, 133
Ishii, S., 84, 85, 113, 133

Jacobson, A. L., 71, 74, 130
Jacobson, R., 74, 133, 138
Janakidevi, K., 106, 112, 130
Jha, R. K., 65, 68, 120, 130
John, E. R., 74, 127
Jones, C. S., 44, 130
Jones, W. C., 18, 19, 20, 130
Josephson, R. K., 41, 65, 110, 116, 120, 130

Kabat, C., 74, 138
Kamada, T., 106, 130
Katz, B., 110, 131
Kidder, G. W., 106, 112, 130
Kimble, D. P., 74, 113
Kirtland, R. J. A., 60, 137
Klein, N., 75, 78, 84, 96, 113, 116, 120, 131
Kleinenberg, N., 5, 131
Koehler, O., 71, 131
Koelle, G. B., 115, 131

Ledbetter, M. C., 60, 131
Lendenfeld, R. von, 20, 44, 131
Lender, T., 75, 78, 84, 96, 113, 116, 120, 131
Lenhoff, H. M., 40, 131
Lentz, T. L., 22, 26, 42, 44, 45, 48, 52, 56, 60, 64, 79, 84, 109, 112, 113, 114, 115, 116, 126, 131, 132, 138
Lesh, G. E., 48, 132
Loewenstein, W. R., 115, 135
Loewi, O., 114, 132
Loomis, W. F., 40, 132
Loubatières, R., 20, 137
Lourie, E. M., 106, 112, 126

McConnell, C. H., 49, 133
McConnell, J. V., 71, 74, 120, 133, 136, 138
McCullough, C. B., 10, 40, 41, 133, 134
Mackay, B., 8, 68, 120, 130
Mackie, G. O., 41, 42, 65, 68, 120, 130, 132
Macklin, M., 41, 65, 116, 120, 130
McNair, G. T., 18, 19, 133
MacRae, E. K., 97, 102, 114, 132
Mariscal, R. N., 45, 132
Mathias, A. P., 53, 113, 132
Maynard, R., 40, 61, 136
Mecklenburg, C. von, 56, 84, 113, 127
Metschnikoff, E., 18, 133
Michaelson, I. A., 60, 137
Mitropolitanskaya, R. L., 26, 53, 112, 113, 133
Moorhead, M., 26, 53, 84, 113, 137
Morita, H., 115, 133
Morita, M., 78, 84, 97, 113, 133
Morowitz, H. J., 43, 125
Müller, H. G., 71, 78, 133
Murbach, L., 44, 133
Myhrberg, H., 56, 84, 113, 127

Nachmansohn, D., 26, 53, 84, 113, 126
Nagai, T., 19, 135
Nakajima, Y., 111, 126
Nystrom, R. A., 19, 135

Olmsted, J. M. D., 74, 133
Oosaki, T., 84, 85, 113, 133
Orton, J. H., 18, 133
Östlund, E., 26, 53, 106, 112, 113, 133

Palay, S. L., 91, 129

Pantin, C. F. A., 8, 9, 10, 12, 44, 108, 109, 110, 114, 118, 127, 134
Pappas, G. D., 111, 126
Pardoe, U., 106, 112, 126
Parker, G. H., 6, 7, 8, 9, 18, 19, 45, 109, 134
Passano, L. M., 8, 9, 10, 40, 41, 109, 110, 118, 134
Pavans de Ceccatty, M., 10, 18, 19, 20, 22, 31, 34, 35, 112, 118, 134, 135, 137
Pedersen, K. J., 4, 135
Pellegrino de Iraldi, A., 60, 91, 128, 135
Picken, L. E. R., 44, 135
Porter, K. R., 60, 129, 131, 136
Potter, D. D., 115, 135
Press, N., 97, 135
Prosser, C. L., 19, 114, 135

Reniers-Decoen, M., 44, 126
Richardson, K. C., 91, 135
Robertson, J. A., 71, 135
Robson, E. A., 43, 114, 127, 135
Röhlich, P., 97, 135
Rosenblueth, A., 116, 135
Ross, D. M., 43, 45, 53, 109, 113, 118, 128, 132, 135, 136
Rushforth, N. B., 40, 43, 61, 136

Salganicoff, L., 60, 128
Schachter, M., 53, 113, 132
Scharrer, B., 111, 114, 115, 136
Scharrer, E., 114, 115, 136
Seaman, G. R., 107, 112, 136
Semal-Van Gansen, P., 42, 136
Shamarina, N. M., 114
Singer, R. H., 40, 136
Skaer, R. J., 44, 135
Sollas, W. J., 6, 136

Spangenberg, D. B., 49, 136
Stewart, C., 20, 136
Sutton, L., 45, 128

Taliaferro, W. H., 71, 136
Taylor, C. V., 106, 136
Thesleff, S., 114, 136
Thompson, R., 74, 120, 136
Tilney, L. G., 60, 129, 136
Török, L. T., 97, 135
Trembley, A., 37, 137
Tuzet, O., 20, 22, 137

Umrath, K., 106, 137

Van Orden, L. S., 91, 137

Warren, M., 45, 126
Welsh, J. H., 12, 26, 53, 84, 111, 113, 137
Wense, T., 106, 107, 112, 125
Werman, R., 114, 137
Westblad, E., 78, 137
Westfall, J. A., 61, 68, 128
Whittaker, V. P., 60, 137
Wiernik, P., 40, 126
Willmer, E. N., 107, 138
Wilson, E. B., 40, 138
Wintermann, G., 18, 138
Wintermann-Kilian, G., 22, 125
Wolfson, C., 106, 138
Wolken, J. J., 97, 138
Wood, J. G., 26, 52, 113, 138
Wood, R. L., 42, 138

Yamashita, S., 115, 133

Zelman, A., 74, 138
Zieher, L. M., 60, 128

SUBJECT INDEX

Aberrant preganglionic neuron, 21
Acetylcholine, 19, 26, 44, 53, 84, 106 f., 112, *112 f.,* 114 f.
Acetylcholinesterase, 22, 23, 26, 52, 53, 60, 79, *80,* 84, 106 f., *112 f.*
Acoel, 4, 70, 121
Acoeloid form, 3 f.
Acontia, 7, 53
Actinia, 53, *112 f.*
Actiniidae, 45
Action potential, 106, 115 f.
Active transport, 107
Adaptation, 2
Adenosine triphosphatase, 44
Adenosine triphosphate, 19, 44
Adjustor, 6, 8
Adrenalin. *See* Epinephrine
Adrenergic neuron, 84, 91
Alcian blue, 23
Alcyonium digitatum, 113
Alleocoela, 70
γ-Amino butyric acid, 19
Amoebocyte, 15, *16,* 31
Ancestral metazoan, 3, 5, 103, *104,* 122
Anthozoa, 36
Arachnoid cell. *See* Spider cell
Archeocyte, 15, *16,* 31
Atrichous isorhiza, 40
Atropine, 19, 45
Aurelia, 53, *112 f.*
Auricle, 71, 78, 102
Axon, 21, 52, 97, *100,* 102, 110. *See also* Neurite; Process
Axoplasmic flow, 60

Behavior, 2, 9, 15, 103–06, 109; hydra, 37–41; planaria, 70–71; sponges, 18–19

Bilateria, 4
Bioluminescence, 106
Bladder cell. *See* Vesiculous neuron
Bouton, 21, 34
Brain. *See* Cephalic ganglion

Calliactis, 45, 56, *113*
Catecholamine, 26, 53, 60, 106, 111, *112 f.,* 115–16
Cell membrane. *See* Plasma membrane
Central nervous system, 120 f., 123
Centriole, 30
Cephalic ganglion, 70 ff., *72,* 84, 120 f.
Cestoda, 69
Chemoreceptor, 70 f., 78, 102
Chemosensitivity, 111, 114, 122
Chemotaxis, 70 f.
Chloroform, 19
Choanocyte, 6, 14, *16,* 18 f., 21
Choanoflagellate, 4
Cholinergic neuron, 60
Chrome hematoxylin, 23
Chromocyte, 15
Cilia, 4, 61, *62,* 69, 102 f., 106 f., 119, 122
Ciliata, 4, 106
Ciliated groove, 70 f., 78
Ciliated pit, 70 f., 78
Citric acid, 40
Classical neuron, 20–21
Clathrina, 20
Cliona, 19 ff.
Cnidoblast, 36, *38,* 44, *46,* 52 f.
Cnidocil, 44
Cocaine, 19
Coelenterata, 1 ff., 15, 36, *112–13,* 120, 123

Collencyte, 6, 15
Commissure, *72,* 75
Conditioned inhibition, 43
Conditioned response, 43, 74, 120
Conditioning, 43, 74
Conduction, 7, 9 ff., 19, 41 f., 65, 106 f., 108, 110–11, 120, 122
Conductor, 5
Contraction, 9, 18 f., 37 f., 41
Contraction–burst system, 41
Copulatory apparatus, 4, 69
Cordylophora, 65, 68
Cornea, 79
Crithidia fasciculata, 112
Crustacean, 110
Cyanea, 53, 68, *113*

Dendrite, 21, 97, *100,* 110
Depolarization, 10, 12, 41, 107, 109 f., 122
Dermal pore, 18 f.
Desmacyte, 15, 18
Desmosome, 35
DFP, 45
Differentiation, 44, 48 f., 108, 119
Digestion, 4 f., 18. *See also* Feeding
Digestive cavity, 36, 41
Digestive cell, 3 f., 36, *38, 103, 104*
Dugesia, 74, 85, 90, 102, *113*

EDTA, 27
Effector, 5 ff., 103, 106 ff., 122 f.; independent, 6 ff., 15, 26, 44 f., 103, 117
Electrical excitability, 111, 120
Electrical inexcitability, 111, 114, 122
Electrical transmission. *See* Transmission, electrical
Electrotonic transmission. *See* Transmission, electrical
Elementary nervous system, 1, 68
Endoplasmic reticulum, 30, 56, 85 ff., 96, 102
Ephaptic transmission. *See* Transmission, electrical
Epidermal cell, 11
Epidermis, 3 f., 14, 36, *38,* 69, 103, *104,* 107
Epinephrine, 19, 22 f., *24,* 44, 52 f., 91, 112. *See also* Catecholamine

Epithelial cell, 5, 7 ff., 36, *76,* 103, *104,* 107, 122
Epitheliomuscular cell, 3 ff., 36, *38,* 42, 52 f., 65, *66*
Epithelium, 14
Eserine. *See* Physostigmine
Ether, 19
Eumetazoa, 4 f., 14, 36. *See also* Metazoa
Euphysa, 65, 68
Evolution, 2 f., 108; of nerve cell, 117–22
Excitation, 11 f., 106 ff., 111, 122
Excretory system, 69, 84
Eye. *See* Ocellus
Eyespot, 106

Feeding, 3, 106; hydra, 40–41; planaria, 70; sponges, 18
Filament, 15, 21, 34 f., 61. *See also* Myofilament
Flagella, 18, 107 f.
Flagellata, 4, 106
Flatworm, 3 f., 69 f., 114, 121. *See also* Planaria; Platyhelminthes
Frontal organ, 70, 78

Galvanotaxis, 71
Gamete, 4, 49
Ganglion, 10, 37, 75, 120. *See also* Cephalic ganglion
Ganglion cell, 7 f., 20, 37, *38,* 40, 49, *50,* 52, 56–57, 60, *82,* 84–85, 118 f., 121
Gastraea theory, 3
Gastrodermis, 36, *38,* 69
Geotaxis, 71
Gland cell, 4 f., 15, 36, *38,* 69, 103, *104,* 122
Glucosamine, 40
Glucose-6 phosphatase, 44
Glucose-6 phosphate, 44
Glutaraldehyde, 27, 91
Glutathione, 19, 40
Glycogen, 5, 7, 64, 96, 102
Golgi apparatus, 30, 57, 85, 90, 97, 119
Grantia, 20 f.
Granule, 23, 27, *28,* 30 f., *32,* 34 f., 57, *88,* 90–96, *92, 98,* 119; neuro-

secretory, 34, 45, 48, *54,* 57, 64, *66,* 68, 91–96, *94, 98*
Granule-containing cell: hydra, *54,* 57–60, 120 f.; planaria, *88,* 91–96, *92,* 121; sponges, *28,* 30–31, *32*
Growth process, 43–44

Habituation, 43
Halichondria, 20 f.
Halisarca, 20
Hexamethonium, 19
Hippopodius, 42
Hippospongia, 22, 35
Histamine, 19, 44
Hormone, 111, 114–16, 118, 123
Hydra, 1, 5, 36–68, 117–20; behavior, 37–41; histology, 36–37, *38;* nervous system, 36–68, *38, 46*
Hydra, 53, *113*
5-Hydroxytryptamine, 19, 22 f., 26, 44, 52 f., *56,* 60, 84, 112, *112 f.*
Hydrozoa, 36, 42, 65
Hymeniacidon. See *Stylotella*

Impulse, 42, 120
Independent effector, 6 ff., 15, 26, 44 f., 103, 117
Internuncial-motor neuron, 117–18
Internuncial neuron, 78, 117
Interstitial cell, *38,* 44, 48 f.

Jellyfish, 36

Learning: coelenterates, 43–44; planaria, 71–74
Leptoplana, 71
Leucandra, 20 ff.
Leucosolenia, 21
Lithocyte, 78
Locomotion, 4, 37–38, 70
Locomotor-perceptive cell, 3
Lophocyte, 22
Lysosome, 57, 85, 90

Magnesium, 7, 19
Mechanoreceptor, 65
Medusa, 36 f., 65
Membrane. *See* Plasma membrane
Mesenchymal cell, 15, 30 ff., *76, 104*

Mesenchyme, 4, 14, 21 ff., 27, 37, 69, *104,* 121
Mesoglea, 36, *38*
Metazoa, 3, 14; origin, 2–5. *See also* Eumetazoa
Methylene blue, *47,* 49 ff., 64 f.; vital staining technique, 49
Metridium, 43, 53, *113*
Microtubule, 34, *50,* 57, 60, 64, *82,* 85, 96 f., 102, 119
Mitochondria, 30, 56, 64, 85 ff., 96 f., 102
Monoamine, 56
Monoamine oxidase, 22 f., 52 f.
Motor cell. *See* Ganglion cell
Motor end plate, 114
Motor neuron, 78, 117
Mucous, 31, 36
Multivesicular body, 30, 90, 97
Muscle, 4 f., 7, 9 f., 69, *76,* 103, 114, 122; smooth, 15, 19, 114, 116; striated, 114. *See also* Myocyte
Mutation, 107
Myocyte, 6, 15, *16,* 18 f., 21, 26, *104,* 114
Myofilament, 15, 34, 36, *66,* 114

Nanomia, 65, 68
Nematocyst, 7, 36, 40; control of discharge, 44–45
Neoblast, 9, *76*
Nerve cell, 2, 5, 10 ff., 109, 118; definition, 12; differentiation, 44, 48, 108–09; evolution, 117–22; identification, 12–13. *See also* specific types
Nerve cord, 4, 70, *72,* 75, 96, 120 f.
Nerve fiber. *See* Axon; Neurite; Process
Nerve net, 4, 7 ff., 42, *46,* 65, 120
Nerve plexus: subepidermal, 75, 79; submuscular, 4, 70, 75, 79
Nerve ring, 10, 37
Nerve termination, 21, 34–35, *46,* 52 f., 64, *66,* 116, 120. *See also* Synapse
Nervous system: central 120 f., 123; definition, 11–13; elementary, 1, 68; hydra, 36–68, *38, 46;* origin, 1–13, 103–23; planaria, 69–102, *72, 76;* primitive, 1–13, 15, *104*

Neurite, *38, 46, 50,* 52 f., 61–64, *66.*
  See also Axon; Process
Neuroglia, 78, 97, 121
Neurohormone, 111–17, *112–13*
Neurohumor, 9, 12, 19 f., 26 f.,
  107, 111–20, *112–13,* 123; coe-
  lenterates, 52–56; hydra, 52–53;
  planaria, 79–84; protozoa, 106–
  07; sponges, 22–26
Neuromuscular cell, 5, 21
Neuromuscular mechanism, 6–9
Neuron. *See* Nerve cell
Neuropil, 75, 84, 96, *98*
Neurosecretion, 96
Neurosecretory cell, 10 f., 45, 49,
  53, *54,* 57, 60, 75, 78, 84, *94,* 111,
  118 f., 123; origin, 10–11, 114–
  15, 118, 123
Neurosecretory granule, 34, 45, 48,
  54, 57, 64, *66,* 68, 91–96, *94, 98*
Neurosecretory substance, 9, 22 f.,
  48 f., 96, 111 f., *112–13,* 118 f.,
  123
Neurosensory cell, *38,* 52, 61, *62,*
  97, 119
Neurotransmitter. *See* Neurohumor;
  Transmitter
Neurotubule. *See* Microtubule
Nicotine, 19
Nicotinic acid, 40
Nissl substance, 21
Noctiluca miliaris, 106, *112*
Norepinephrine, 22 f., *24,* 44, 52 f.,
  91, 112. *See also* Catecholamine
Nucleolus, 30, 56, 84 ff., 97
Nucleus, 21, 30, 56 f., 84 ff., 97
Nutritive cell, 3, 36. *See also* Diges-
  tive cell

Ocellus, 37, 65, 70, 78 f.
Olfactory pit, 37
Osculum, 14, 18 f., 22
Osmium, 20, 27, 91
Ostium, 14, *16,* 18

Pacemaker, 10, 41
Pachymatisma, 20 f.
Paraldehyde fuchsin, 53, 84
Paramecium, 112
Parazoa, 3 f., 14

Periodic acid Schiff, 53
Peristaltic waves, 7, 70
Permeability, 12, 107, 109
Phagocytic vacuole, 30 f.
Photoreceptor, 61, 97, *100,* 121
Phototaxis, 71
Physalia, 56
Physostigmine, 19, 45
Pigment cell, *5,* 31, 69, 79, 97, 103,
  104
Pinacocyte, 6, 14, *16,* 18, 21, 35
Planaria, 1, 69–102, 110, 120–21;
  behavior, 70–71; histology, 69–
  70, *76;* nervous system, 69–102,
  72, 76, 120–21
Planaria, 113
Planula, 3 f.
Planuloid form, 3 f.
Plasma membrane, 12, 34, 56, 64,
  97, 106 f., 109
Platyhelminthes, 1, 3, 69, *113,* 123
Polycelis nigra, 75, 84, *113*
Polyorchis penicillatus, 61, 68
Polyp, 36 f., 110
Porifera, 1, 3, 14, *112,* 123. *See also*
  Sponge
Porocyte, 14
Potassium, 19, 109
Potential, 12, 41 f., 65, 106, 110,
  115 ff.
Prenervous cell, 2, 107, 112, 122
Primitive nerve cell: nature, 109–
  17; origin, 107–09
Primitive nervous system, 1–13, 15,
  104. See also Primitive nerve cell
Process, 21 ff., 27, 30, 52, 61–64,
  76, 78 ff., *80, 82,* 107. *See also*
  Axon; Neurite
Procotyla, 79, 84, 90, *113*
Protista, 3
Protomyocyte, 9
Protoneuron. *See* Ganglion cell
Protozoa, 4, 15, 103, 106–07, 110,
  112

Receptor, 5 ff., 40, 49, 61, 78 f.,
  106, 109. *See also* Sensory cell
Reflex, 9 f.
Regeneration: hydra, 44–49, 60;
  planaria, 74–75, 120

Reserpine, 45
Response, 7, 18 f., 26, 37 f., 42, 71 f., 109 f., 115; all-or-none, 12, 106, 110–11, 117, 123; graded, 12, 110–11, 122
Reproduction, 5
Reproductive cell, 3
Reproductive system, 69
Rhabdite, 69, *76*
Rhabdoid, 4, 69
Rhabdome, 97, *100*
Rhabdomere, 121
Rheoreceptor, 70, 78
Rheotaxis, 71
Rhizostoma pulmo, 113
Rhythmic potential system, 41
Ribosome, 30, 56, 64, 85 ff., 96, 102
RNA, 74

Sagartia, 53, *113*
Sagittocyst, 4
Sarsia, 65, 68
Scleroblast, 15, *16*, 31, 34
Scyphomedusae, 36 f.
Scyphozoa, 36
Sea anemone, 7, 26, 36, 45, 53 f., 109 f., 118
Secretion, 11, 107 f., 112, 122
Secretory cell, 10, 11, 108, 114
Selection, 2, 103
Sensorimotor cell, 109, 117
Sensory cell, 4, 6f., 10, 20, 22, 37, *38*, 40, 49, 52 f., *58*, 60–61, 75, *76*, 78, 109, 117, 119, 121, 123
Sensory hair, 5, 52, *58*, 65–68, 78
Sensory margin, 78
Sensory organ, 10, 71, 78, 120 f.
Serotonin. *See* 5-Hydroxytryptamine
Silver impregnation, 20, 31
Sodium, 109
Somatic cell, 3
Sphincter, 6, 8, 19
Spicule, 14 f., *16*, 22, 27
Spider cell, 21
Spike. *See* Depolarization; Response, all-or-none
Sponge, 1, 3, 6 f., 14–35, 118–19; behavior, 18–19; histology, 14–15, *16;* question of a nervous system, 15–18, 27, 35, 118–19. *See also* Porifera
Spongin fiber, 15
Spongocoel, 14
Statocyst, 37, 70 f., 78
Statolith, 78
Stimulus, 6 f., 12, 18 f., 40, 43, 109–10, 115, 122
Stoichactiidae, 45
Stomphia coccinea, 43
Strychnine, 19
Stylotella, 7, 19
Subepidermal nerve plexus, 70, 75, *76*, 78 f., *80*
Submuscular nerve plexus, 4, 70, 72, 75, *76*, 78 f., *80*
Sycon, 20, 22, 27, *112*
Synapse, 8, 35, 42, 64 f., 68, 97, *98*, *100*, 102, 111, 116, 120 f., 123
Synaptic vesicle, 60, 90, 102, 115
Synocil, 20

Tactile receptor, 70, 78
Tealia felina, *113*
Tethya, 19, 31, *112*
Tetrahymena, *112*
Thermotaxis, 71
Thesocyte, 15
Thigmotaxis, 71
Threshold, 12, 45, 116
Through-conduction, 10, 41, 65
Toluidine blue, 53
Transmission, 7 f., 12, 34, 107 f., 112, 114, 120, 122 f.; chemical, 12, 65, 108, 111 f.; electrical, 12, 65, 111
Transmitter, 12, 114 ff. *See also* Neurohumor
Trematoda, 69
Trophic effect, 117, 123
Trypanosoma rhodesiense, *112*
Tryptamine, 19
D-Tubocurarine, 19
Tubularia, 53, *113*
Turbellaria, 69 f., *113*, 121

Undifferentiated cell, 4, 69, 103, *104*. *See also* Archeocyte; Interstitial cell; Neoblast
Urea, 40

Vesicle, 23, 30, 34 f., 57, 60, 68, *82,* 85, *86,* 96, *98, 100,* 102, 119; synaptic, 60, 90, 102, 115

Vesiculous neuron, 21
Vital staining, 49
Volvox, 3